ORDER NUMBER EA-AGV

AIRCRAFT GOVERNORS

An IAP, Inc.
Training Manual

By Frank Delp

International Standard Book Number 0-89100-156-5
For sale by: IAP, Inc.
Mail to: P.O. Box 10000, Casper, WY 82602-1000
Ship to: 7383 6WN Road, Casper, WY 82604-1835
(800) 443-9250 • (307) 266-3838 • FAX: 307-472-5106
HBC0990 Printed in the USA

IAP, Inc.
7383 6WN Road, Casper, WY 82604-1835

TABLE OF CONTENTS

Preface

This book on *Aircraft Governors* is one of a series of specialized training manuals prepared for aviation maintenance personnel.

This series is part of a programmed learning course developed and produced by International Aviation Publishers (IAP), one of the largest suppliers of aviation maintenance training materials in the world. This program is part of a continuing effort to improve the quality of education for aviation mechanics throughout the world.

This manual is designed to present the A&P mechanic/student with the information necessary for a general understanding of the theory, operation, and maintenance of aircraft governors.

The information contained in this manual is for instructional purposes only and is not to be used as a substitute for a manufacturer's current maintenance manual, service bulletins, or operational data.

This particular manual on *Aircraft Governors* includes a series of carefully prepared questions and answers to emphasize key elements of the study, and to encourage you to continually test yourself for accuracy and retention as you use this book.

Acknowledgements

The validity of any program such as this is enhanced immeasurably by the cooperation shown IAP by recognized experts in the field, and by the willingness of the various manufacturers to share their literature and answer countless questions in the preparation of these programs.

We would like to mention, especially, our appreciation for help given us by:

McCauley Accessory Division
of Cessna Aircraft
Hartzell Propeller, Inc.
Woodward Governor Company
Pratt & Whitney Aircraft of Canada
Hamilton Standard
(Division of United Technologies)

If you have any questions or comments regarding this manual, or any of the many other textbooks offered by IAP, simply contact: Sales Department, IAP, Inc.; Mailing Address: P.O. Box 10000, Casper, WY 82602-1000; Shipping Address: 7383 6WN Road, Casper, WY 82604-1835; or call toll free: (800) 443-9250; International, call: (307) 266-3838.

SECTION I

History of Governors

Early in the development of the airplane it became apparent that many factors had to be taken into account to improve aircraft performance. One of the factors was the propeller design in relation to the most efficient blade angle. The first propellers were fixed pitch and if the propeller blade angle is set to give good takeoff and climb performance, the propeller would be inefficient in cruising flight because the blade would be at too low an angle. If the propeller blade angle was set for the best cruise performance, the takeoff run would be relatively long with a poor rate of climb for the aircraft.

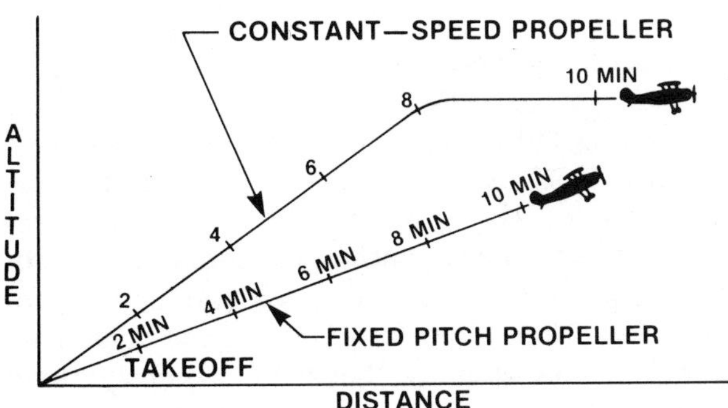

Fig. 1-1 A comparison of the performance of an aircraft equipped with a fixed-pitch propeller versus an aircraft equipped with a constant-speed propeller.

A constant-speed propeller was needed along with a device that would change the propeller blade angles with varying flight conditions to obtain an acceptable amount of efficiency during all flight conditions. This device was the propeller governor.

The propeller governor can sense the speed (RPM) of the engine and adjust the propeller blade angle to maintain a selected RPM regardless of the operational conditions of the aircraft. This allows a low blade angle during takeoff to obtain maximum RPM and engine power and an increase in blade angle as the aircraft accelerates to prevent overspeeding of the engine.

A. A Brief History of the Propeller Governor

In the early 1930's the Hamilton-Standard Company worked with the Woodward Governor Company to develop a governor that would control the hydraulically operated propeller developed by Hamilton-Standard to give automatic engine speed control.

As engines and aircraft increased in size and power during the 1930's and 1940's, governor systems were refined to give more precise control and increase the governor system capabilities to allow feathering and reversing of the propeller blades.

During the early development of the governor systems most emphasis was directed toward transport category with relatively little attention given to light aircraft governor systems until after World War II. However, the transport designs provided the technology to develop the

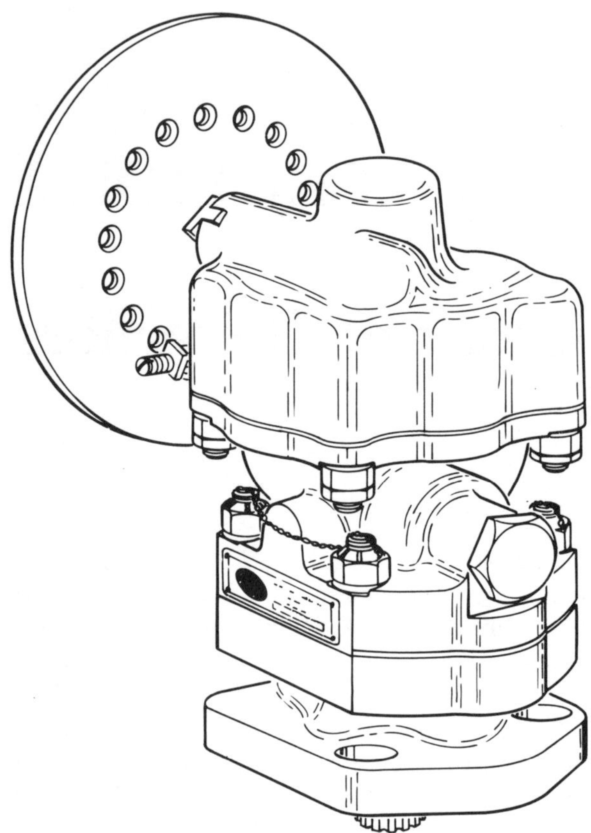

Fig. 1-2 An early Woodward governor used with Hamilton-Standard propellers.

COURTESY OF DOUGLAS AIRCRAFT CO.

Fig. 1-3 The Douglas DC-3 with constant-speed and feathering propeller systems.

smaller, lighter, and less complex governor designs needed for the smaller engines and propellers of light aircraft.

One of the first light aircraft to be fitted with a constant-speed system was a 1946 Culver Cadet equipped with a 75 HP engine, a Sensenich variable pitch propeller, and a Woodward governor. This aircraft provided the basis for modifications to adapt constant-speed systems to other aircraft.

The first use of propeller constant-speed systems on modern light aircraft was on such designs as the Beech Bonanza and the Cessna 180 in the late 1940's and early 1950's. These were some of the first light aircraft which used horizontally opposed engines designed with a governor drive pad and internal oil transfer seals to transfer oil pressure from the governor on the engine case to the propeller on the crankshaft. As more engines were designed with these features, constant-speed systems came into wide usage in the light aircraft industry.

B. Governor Manufacturers

There are three propeller governor manufacturers who supply the majority of aircraft governors used in the United States: McCauley, Hartzell, and Woodward.

1. McCauley governors

The McCauley Accessory Division of Cessna Aircraft Company of Vandalia, Ohio, now supplies most of the governors, unfeathering accumulators, synchronization systems, and propellers used on Cessna aircraft and some propellers for other aircraft manufacturers such as Beechcraft and Mooney. These governor systems were previously supplied by Woodward.

2. Hartzell governors

The Hartzell Propeller Corporation of Piqua, Ohio, supplies most of the original equipment for the constant-speed systems on Piper aircraft and many of the propellers for Beech aircraft. The corporation also has systems which may be used on aircraft of other manufacturers either as original equipment or through Supplemental Type Certificates.

COURTESY OF BEECH AIRCRAFT CORP.

Fig. 1-4 The Beech Bonanza was one of the early production light aircraft to offer a constant-speed system.

Most Hartzell governors are war surplus governors which have been reworked to produce the desired operations.

3. Woodward governors

The Woodward Governor Company of Rockford, Illinois, provides most of the governors used on light turboprop powered aircraft and governors which are approved through Supplemental Type Certificates for many other aircraft. Woodward also provides many of the governor systems (fuel controls) used with transport aircraft and turbojet and turbofan powered aircraft. Woodward governors are used on many Beech aircraft and several Piper and Cessna models.

QUESTIONS:

1. What does a governor "sense"?
2. What was one of the first light aircraft to use a governor system?
3. What engine design changes were necessary to allow constant-speed operation?
4. Which company provides most of the governors used on Piper aircraft?

SECTION II

Basic Governor Construction

A propeller governor must be driven directly or indirectly from the engine crankshaft so that it can sense the engine RPM. This may be above or below the crankshaft RPM, but it must be a fixed proportion to the engine RPM. The governor must also be able to accept oil from the engine and increase the pressure to a value that will cause rapid propeller response to maintain a constant system RPM.

The governor must be designed to direct the oil pressure to or from the propeller so the propeller blade angles can be changed. It must also be constructed so that the pilot can adjust it in flight to select the desired RPM setting.

A. Governor Components

1. Housing

The governor housing is usually cast from aluminum and may be composed of three primary sections or of one basic section.

a. Three primary sections

If the housing is made of three sections, they will be of approximately equal size. The *head* which contains the pilot's control components, will be at the top of the governor. The *body*, in the middle, contains the sensing components and oil pump plus components which direct the oil flow to and from the propeller and contains the pressure regulation unit. The *base* contains the mounting surface which mates the governor to the engine drive pad.

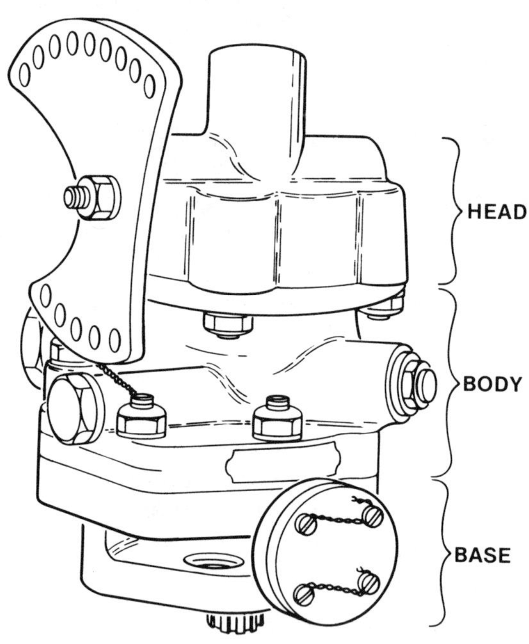

Fig. 2-1 An early Woodward governor with three distinct housing sections.

b. One basic section

More modern designs contain all of the components in the body with only a cover plate used on the top and the bottom of the governor.

Regardless of the design, the housing will be machined to close tolerances to minimize oil leakage between the housing and the internal operating components. There are no seals or

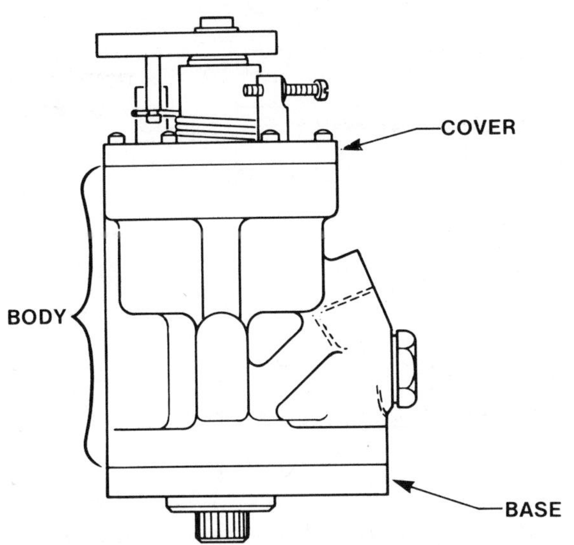

Fig. 2-2 A modern McCauley governor with only one primary housing section and two end covers.

gaskets used inside the governor to prevent oil leakage, except between the housing components.

Some governors have oil passages to allow operation in either a clockwise or counterclockwise direction, depending on the engine drive train and the position of plugs in the governor oil passages.

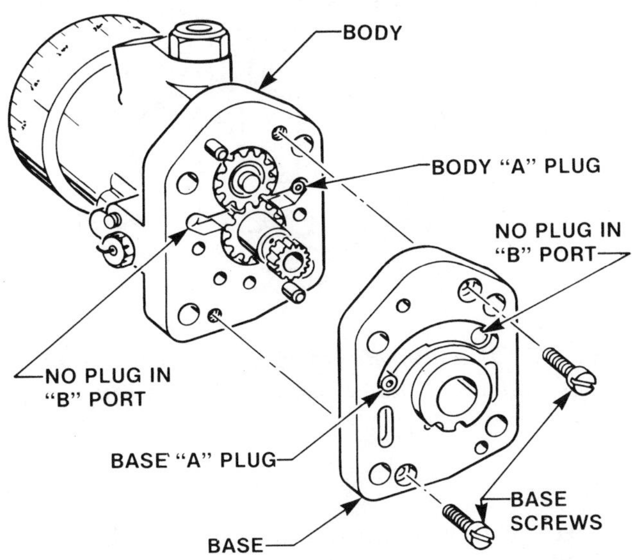

Fig. 2-3 The oil passages in this Woodward governor can be changed to allow the governor to operate with either clockwise or counterclockwise rotation.

2. Drive shaft

The governor drive shaft mates with the engine drive gear on the governor mounting pad. The shaft is used to develop governor oil pressure and give governor components a rotational speed proportional to that of the engine allowing the engine RPM to be sensed. Depending on the engine, the governor may rotate between 80% and 110% of the engine RPM.

The drive shaft has one of the two oil pump gears attached to it or machined as a part of it. The second oil pump gear is driven by the drive shaft gear and is located on a hollow steel shaft.

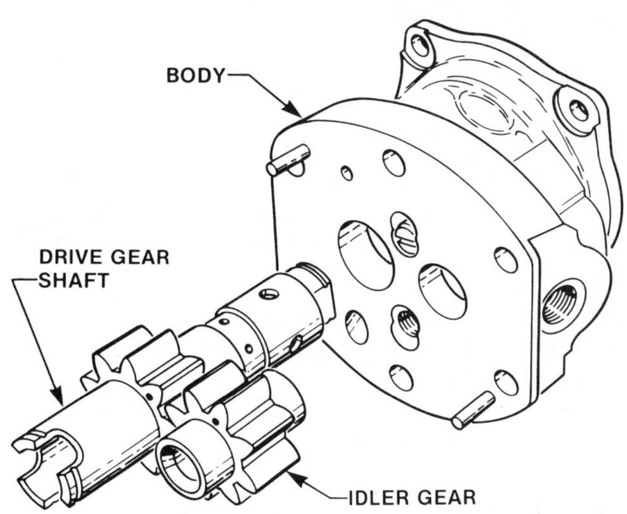

Fig. 2-4 A typical governor drive shaft and oil pump gears.

The oil pump receives oil from the engine lubrication system at engine oil pressure and increases the pressure to two or three hundred psi so that the propeller will operate properly.

3. Relief valve

Oil pressure generated by the governor oil pump is directed through passages in the governor to a pressure relief valve. This relief valve consists of a plunger and a spring, and maintains a constant system operating pressure. When oil pressure reaches the setting of the relief valve, the plunger will off-seat and release excess pressure to the inlet side of the governor pump through the hollow oil pump idler gear shaft, thus maintaining a constant oil pressure

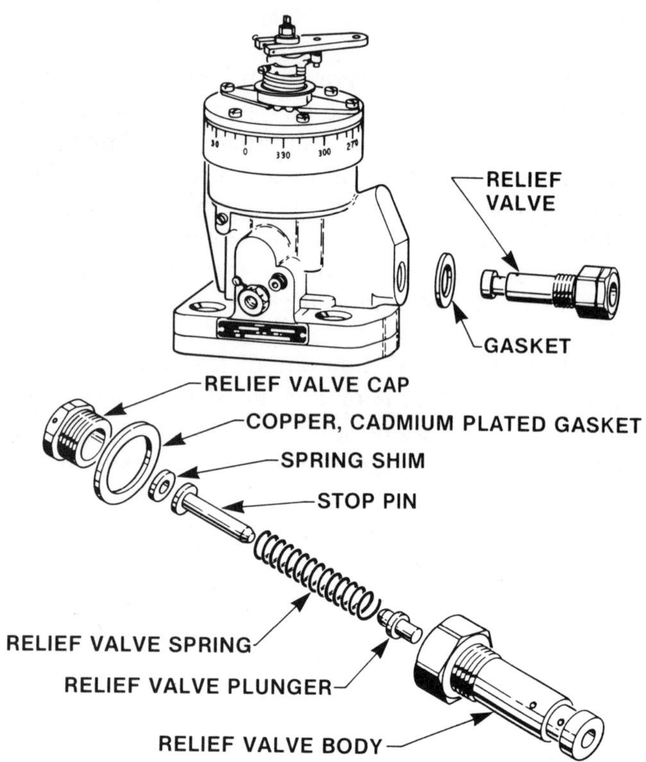

Fig. 2-5 The governor relief valve is used to keep a constant governor oil pressure regardless of the governor RPM.

4. Pilot valve

The regulated governor oil pressure is directed to the propeller, or released from the propeller, as necessary to maintain the selected system RPM by the movement of the pilot valve.

The pilot valve is located in the center of the governor drive shaft, but does not rotate with the shaft. The drive shaft has oil ports machined on it that line up with the oil passages in the governor housing. The pilot valve receives oil through these ports and directs the oil to other drive shaft ports where it flows to, or is released from, the propeller to change the blade angles.

5. Flyweights

The governor flyweights are attached to the top of the drive shaft in the head of the governor and are used to sense the RPM of the engine. The lower part of the flyweights, called the toe, supports the pilot valve. The top of the flyweight contains the weight which reacts to the centrifugal force created by rotation and is used to raise the pilot valve as RPM increases.

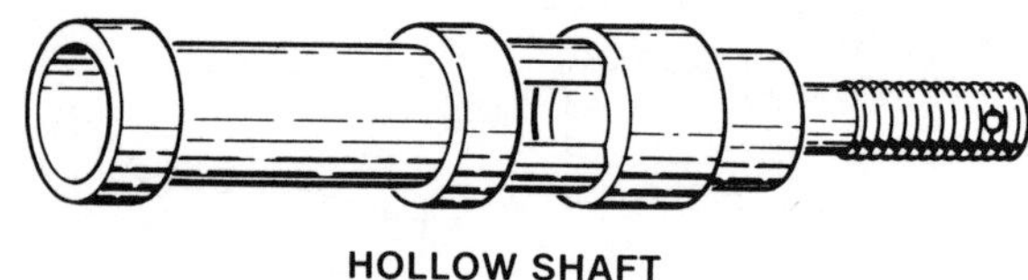

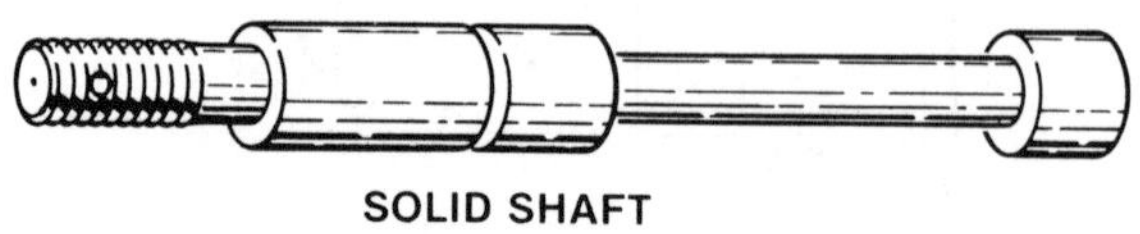

Fig. 2-6 The pilot valve may be one of two designs—a hollow shaft or a solid shaft.

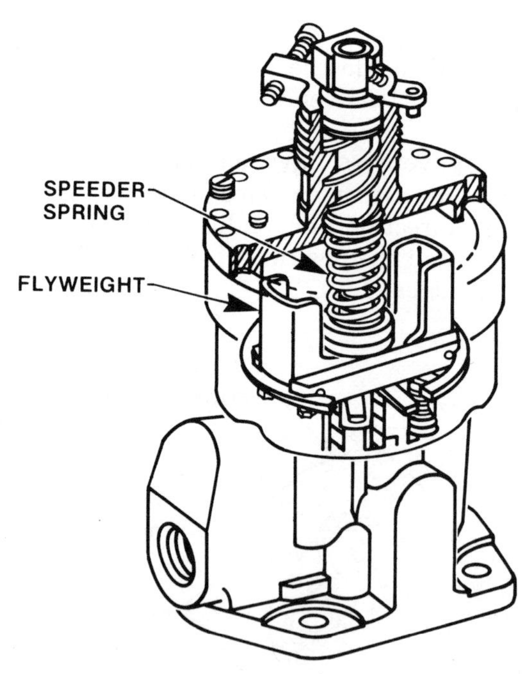

Fig. 2-7 The flyweights respond to drive shaft RPM and are usually located on the top of the governor.

If the flyweights tilt inward, the pilot valve will be lowered and if they tilt outward, the pilot valve will be raised. A ball or roller bearing assembly is used between the toe of the flyweights and the pilot valve so that the flyweights can rotate while the pilot valve does not rotate.

6. Speeder spring

The movement of the flyweights tilting outwards by centrifugal force is opposed by a speeder spring located above the pilot valve and between the flyweights. The compression of the speeder spring is adjusted by the pilot through the cabin propeller control to change the RPM setting of the governor.

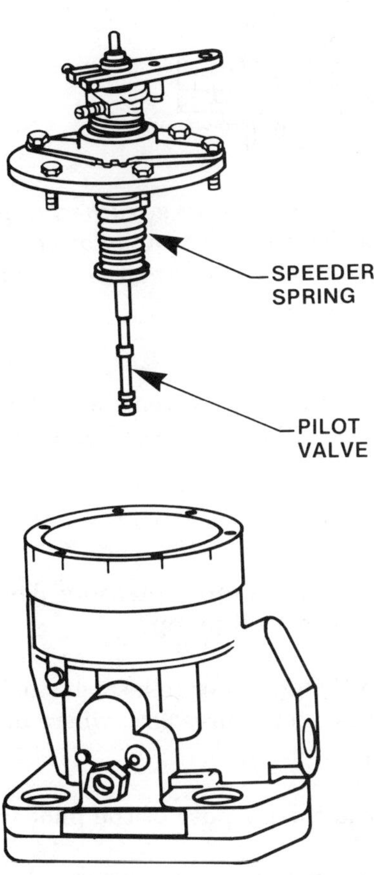

Fig. 2-8 The speeder spring compression is set by the pilot to set the desired engine RPM.

The flyweights act to raise the pilot valve and the speeder spring exerts a force to lower the pilot valve.

7. Speeder rack

The speeder rack (also called the speeder shaft) is located above the speeder spring and is a spiral shaft, or a rack and pinion mechanism, that allows the pilot to adjust the RPM setting of the governor by adjusting the compression of the speeder spring. More compression is achieved when the cabin control is moved forward and the speeder rack sets the governor for a higher RPM. When the cabin control is pulled back, the speeder rack is raised and the compression of the speeder spring is reduced.

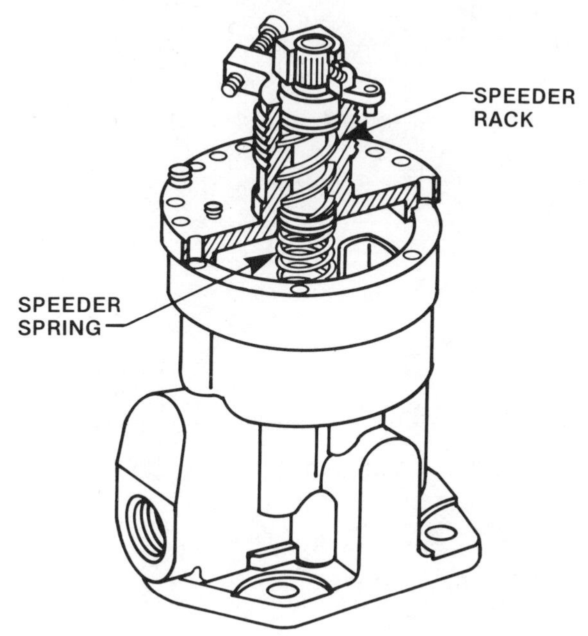

Fig. 2-9 The speeder rack is used to set the compression of the speeder spring.

B. Feathering Governor Components

A feathering governor contains all of the components of a constant-speed governor plus additional components which are necessary to override the constant-speed operation to allow feathering and, in some models, a mechanism to aid in unfeathering the propeller.

1. Lift rod

The lift rod in a feathering governor is located on the speeder rack and extends down into the pilot valve. When the cabin propeller control is pulled aft to the feather position, the speeder rack is raised and raises the lift rod. The lift rod will contact the pilot valve and lift it to cause the propeller to feather.

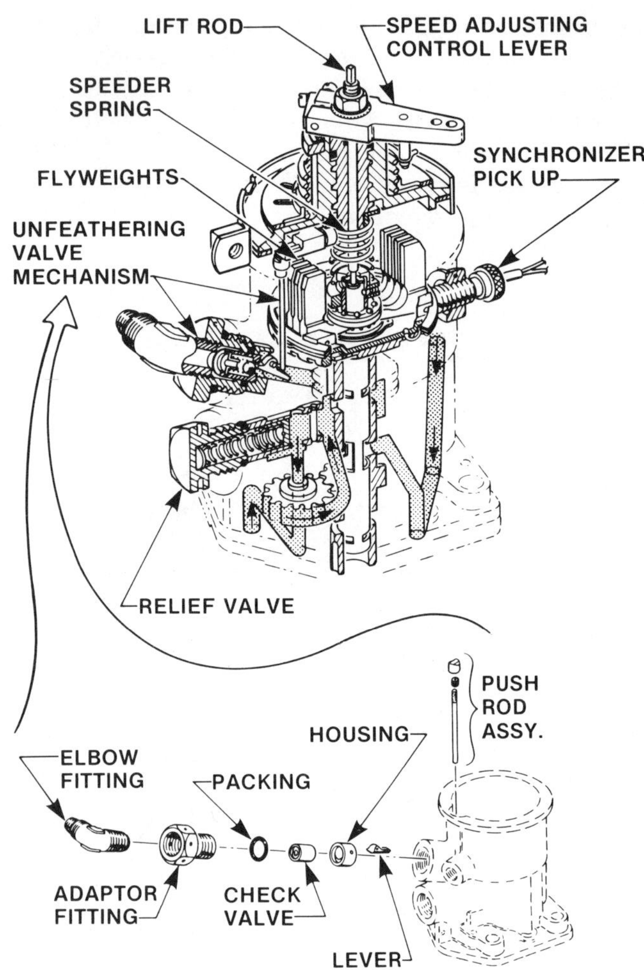

Fig. 2-10 The check valve is used to trap oil in the accumulator when the propeller is feathered.

During constant-speed operation of the governor, the lift rod extends below the lifting surface on the pilot valve and does not interfere with the normal pilot valve movement.

2. Unfeathering check valve

In some governors, a check valve is used to connect an unfeathering oil line to the governor and only allow oil to flow to or from the governor and propeller to allow unfeathering. The specifics of the check valve varies, depending on the type of unfeathering system used.

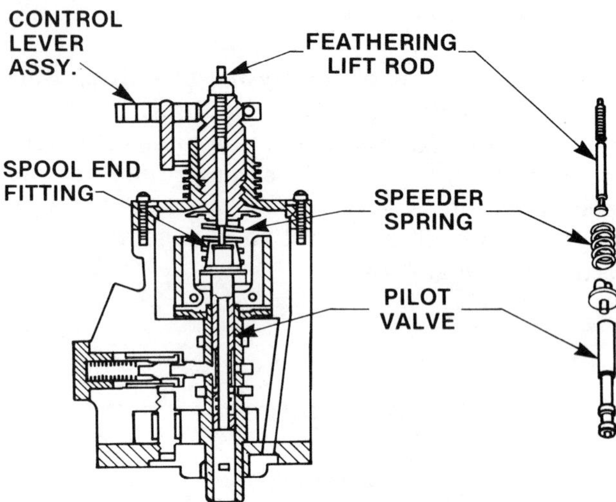

Fig. 2-11 The lift rod in a feathering governor is used to mechanically lift the pilot valve and override the governor constant-speed operation.

The unfeathering systems and the associated check valves will be discussed in a later section.

QUESTIONS:

1. What is the normal pressure developed by the governor oil pump?
2. When the governor relief valve off-seats and releases excess pressure, where does the excess pressure go?
3. What is the purpose of the pilot valve?
4. If the flyweights tilt inward, what happens to the pilot valve?
5. Which governor component opposes the tilting outward of the flyweights?
6. What movement of the cabin propeller control will cause the compression on the speeder spring to decrease?
7. What is the purpose of a lift rod?

SECTION III

Constant-Speed Systems

A. Constant-Speed Operations

There are three basic governor conditions which can exist during constant-speed operation to maintain the engine at a constant RPM. They are: onspeed, overspeed, and underspeed conditions.

1. Onspeed condition

If the engine is operating at the RPM set on the governor by the pilot, the system is said to be onspeed. During the onspeed condition the force of the speeder spring is balanced against the centrifugal force acting on the flyweights to position the pilot valve so as to block the oil passage between the governor and the propeller and hold the propeller at the existing blade angle.

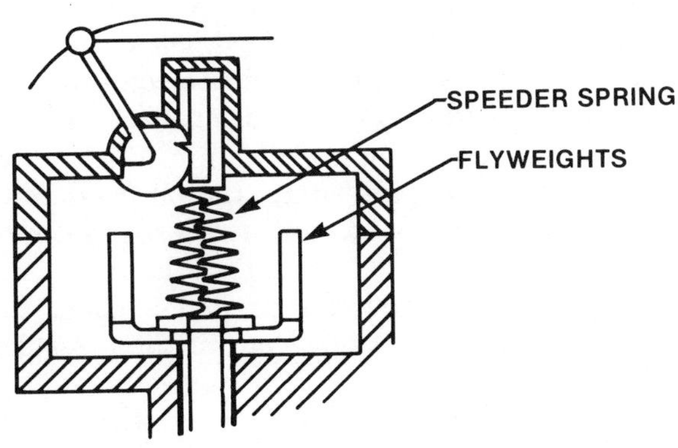

Fig. 3-1 When the governor is onspeed the flyweights and speeder spring are balanced.

2. Overspeed condition

If the engine RPM increases during constant-speed operation, the governor will go into an overspeed condition to return the engine to the selected RPM.

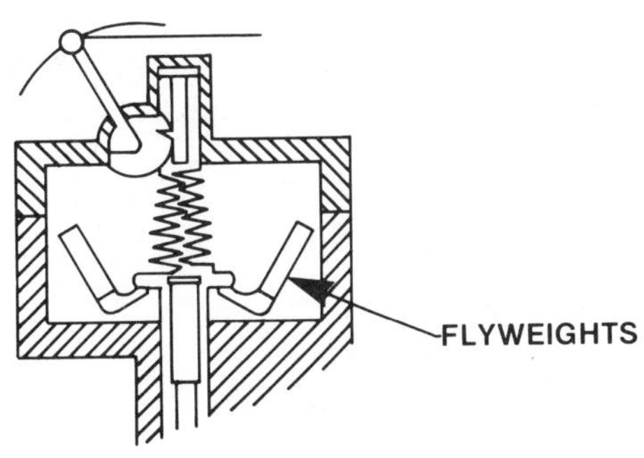

Fig. 3-2 When the governor is overspeed the flyweights tilt outward to raise the pilot valve.

When the governor is in the overspeed condition, the higher RPM causes the centrifugal force on the flyweights to increase and tilt the flyweights outward. As the flyweights tilt outward, the movement is opposed by the speeder spring and the spring is compressed. As the flyweights tilt outward, the toe of the weights lift the pilot valve in the governor to direct the oil flow to cause the propeller blade angle to increase.

As the blade angle increases, the resistance to rotation increases and the engine RPM will slow down. With the reduced engine speed, the flyweights will have less centrifugal force acting on them and the flyweights will be forced to tilt inward by the speeder spring pushing on the toe of the flyweights. As the flyweights tilt inward, the pilot valve is lowered and the flow of oil will be reduced.

This action will continue until the pilot valve returns to the onspeed position and all oil flow to change the propeller blade angle ceases. The propeller blade angle will now remain fixed as long as the system RPM does not change.

The overspeed condition can exist if the throttle is pushed in to increase engine power, if the airspeed increases when the aircraft is placed in a dive, or when the propeller cabin control is moved rearward to set the system for a lower RPM.

3. Underspeed condition

If the system RPM is below the setting of the governor, the governor mechanism will react directing oil flow to or from the propeller as necessary to decrease the propeller blade and return the system to the onspeed condition.

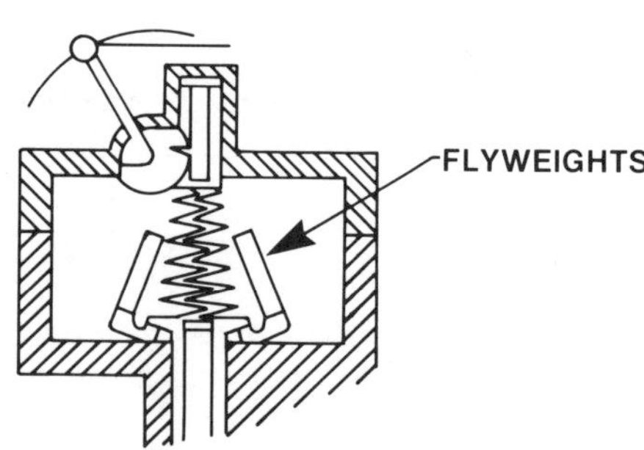

Fig. 3-3 When the governor is underspeed the flyweights tilt in to lower the pilot valve.

When the governor is underspeed, the centrifugal force on the flyweights decreases and the force of the speeder spring will force the flyweights to tilt inward and lower the pilot valve in the driveshaft. The lowered pilot valve will cause the oil in the system to flow to or from the propeller as necessary to decrease the propeller blade angle.

The decrease in the blade angle will reduce the resistance to rotation and the system RPM will increase until the centrifugal force on the flyweights increases to again equal the force of the speeder spring, raise the pilot valve, and return the governor to the onspeed condition.

The underspeed condition can occur when the engine power or aircraft speed is reduced, or when the governor is set for a higher RPM by moving the cabin propeller control forward.

B. Propeller Operation

The design and operational characteristics of the light aircraft propeller can be illustrated by separating the propellers into two types: counterweighted (fixed force to increase pitch) propellers and non-counterweighted (fixed force to decrease pitch) propellers. A complete discussion of propeller operations is not appropriate for this text and only the basic principles of operations will be presented here. (It should be noted that there are exceptions to this operational classification depending upon the purpose of the counterweights.)

1. Counterweight (fixed force to increase pitch) propellers

Counterweighted propellers are those propellers which use weights attached to the inboard section of the blades as the fixed force to increase the propeller blade angle. The oil pressure generated by the governor is used as the variable force to decrease the propeller blade angle.

When the governor is in an overspeed condition and the pilot valve is raised, the oil passage in the governor is opened to release some oil from the propeller and allow it to drain back through the governor to the engine oil sump. With the oil line open to the engine sump, the oil pressure in the propeller is released and the centrifugal force on the counterweights, created by the rotation of the propeller, will move the propeller blades to a higher angle as the oil flows out of the propeller.

As the system RPM decreases with the increase in blade angle, the pilot valve returns to the onspeed position and the remaining oil in the propeller is trapped and the blade angle is again held constant at a higher angle.

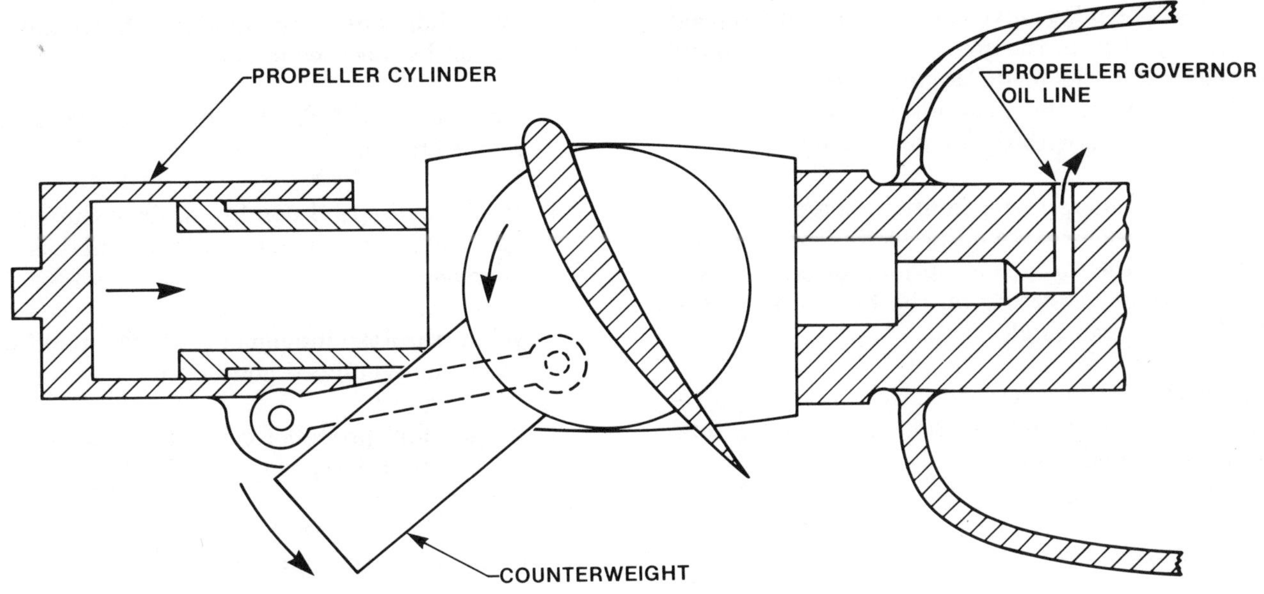

Fig. 3-4 A counterweighted propeller uses the centrifugal force on the counterweights to increase the propeller blade angle.

When the system goes into an underspeed condition, the pilot valve in the governor lowers to direct oil pressure to the propeller to decrease the blade angle. The oil pressure applied to the propeller overcomes the centrifugal force on the counterweights to cause the blade angle decrease.

As the system RPM increases, the pilot valve is raised to the onspeed condition and the oil in the propeller is again blocked, holding the blades at a constant angle.

2. *Non-counterweight (fixed force to decrease pitch) propellers*

Non-counterweighted propellers use governor oil pressure to increase the propeller blade angle and a fixed force such as a spring, or an operational force such as centrifugal twisting moment (CTM) to decrease the blade angle.

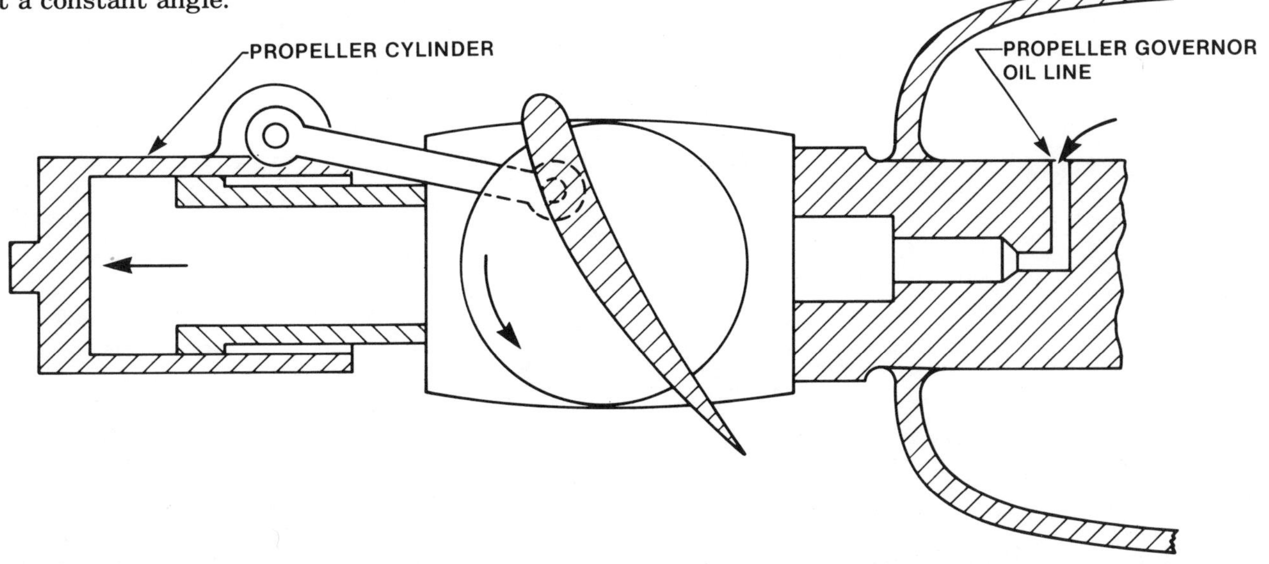

Fig. 3-5 A non-counterweighted propeller uses oil pressure to increase the propeller blade angles.

When the pilot valve is raised in an overspeed condition, oil from the governor is directed to the propeller and the blade angle increases. As the angle increases, the system returns to the onspeed condition, with the blades at a higher angle.

In the underspeed condition, the pilot valve lowers and releases the oil pressure from the propeller and the propeller spring, or centrifugal twisting moment acting on the blades move the blades to a lower angle.

As the RPM decreases, the pilot valve returns to the onspeed position and the oil flow from the propeller stops.

QUESTIONS:

1. What is the relationship of the forces on the flyweights and the speeder spring when the governor is onspeed?
2. What happens to the governor flyweights in the underspeed condition?
3. On which part of the flyweights does the pilot valve ride?
4. In the overspeed condition, does the governor cause the blade angle to increase or decrease?
5. What throttle movement by the pilot will cause an overspeed condition?
6. What cabin propeller control movement will create an underspeed condition?
7. What blade angle change is caused by centrifugal twisting movement?
8. If the pilot valve is lowered in a system using a non-counterweighted propeller, does oil flow to or from the propeller?

SECTION IV

Feathering Systems

Propeller feathering systems are used on most multi-engine aircraft to reduce the drag created by a windmilling propeller if an engine should fail. To be able to feather, the propeller must have a great enough blade angle range to achieve an approximate 90° blade angle. Also, the governor must be designed so that the constant-speed operation can be overidden at anytime to feather the propeller.

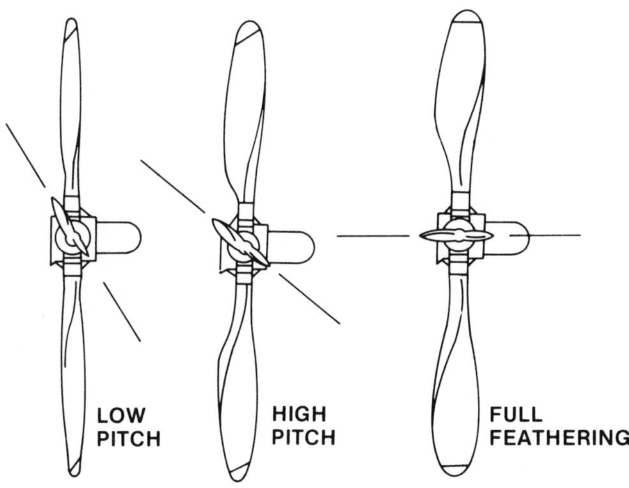

Fig. 4-1 The feather angle is 90° to the plane of rotation whereas the constant-speed blade angle is between 10 and 25° to the plane of rotation.

A means of unfeathering in flight is required in the system and this may require additional system components. These additional components may include accumulators, electric oil pumps, and oil flow control valves.

A. *Basic Feathering System*

The basic constant-speed operating components of a feathering governor are the same as those used in a constant-speed governor. In most feathering governors, the pilot valve and the governor oil passages are designed so that oil pressure is directed to the propeller to decrease the propeller blade angle (underspeed condition) and oil pressure is released from the propeller to increase the blade angles (overspeed condition).

The primary design difference between a constant-speed and a feathering governor is the addition of a lift rod connected to the pilot valve to mechanically lift the pilot valve to the overspeed position when it is desired to feather the propeller. The lift rod raises the pilot valve when the cabin propeller control is moved full aft. When the pilot valve is raised, all of the oil pressure in the propeller is released and the blades move to the feather angle by the centrifugal force on the blade counterweights and some other force, such as springs, depending on the propeller design.

The full aft movement of the cabin propeller control usually requires additional effort or a sideways or outward movement to enter the feather position. This is to prevent accidental feathering of the propellers during constant-speed adjustment.

When unfeathering, the cabin propeller control is moved forward into the constant-speed range. This lowers the lift rod far enough so that it disengages the pilot valve and the governor returns to constant-speed operation (underspeed

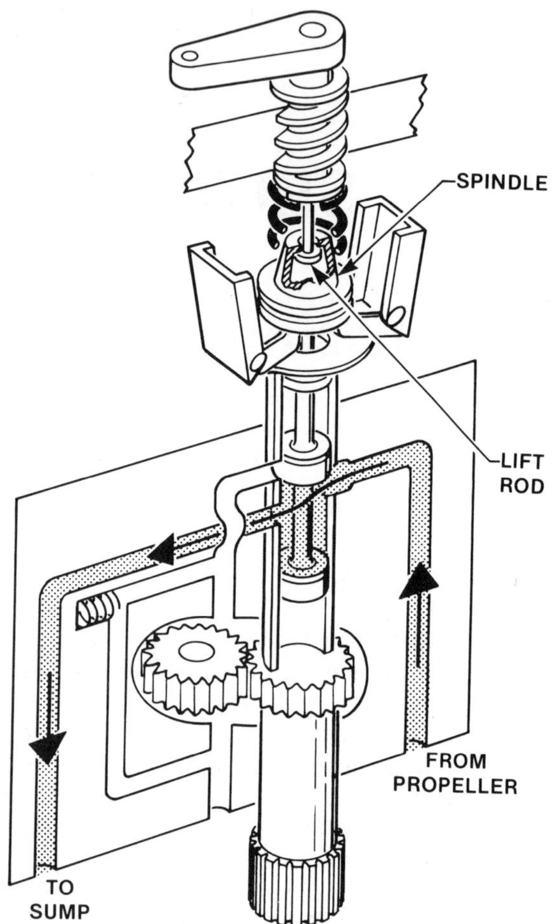

Fig. 4-2 The lift rod engages the pilot valve to feather the propeller.

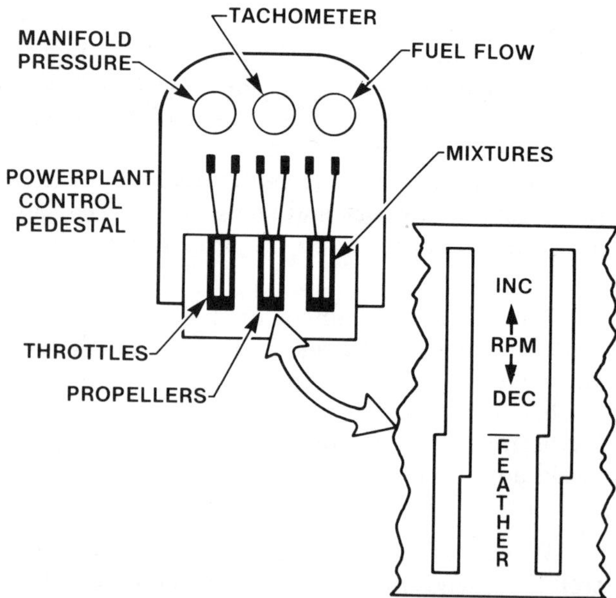

Fig. 4-3 The cabin propeller control may have to be moved sideways to go into the feather range on the control quadrant.

condition before the engine restarts) and allows the engine to be restarted.

The engine is usually restarted by supplying oil pressure to the propeller through the governor, which causes the blade angle to decrease and the engine to windmill. Once the engine is windmilling, the ignition and fuel for the engine is turned on according to the particular aircraft operations manual and the system returns to normal operation.

B. Feathering Propellers

The propellers used with feathering systems are basically the same as constant-speed propellers except for the additional blade angle range and a longer nose cylinder to accommodate the greater blade angle range. These propellers use oil pressure from the governor to decrease the propeller blade angles and counterweights, springs, and air pressure in various combinations to increase the blade angles and feather the propellers.

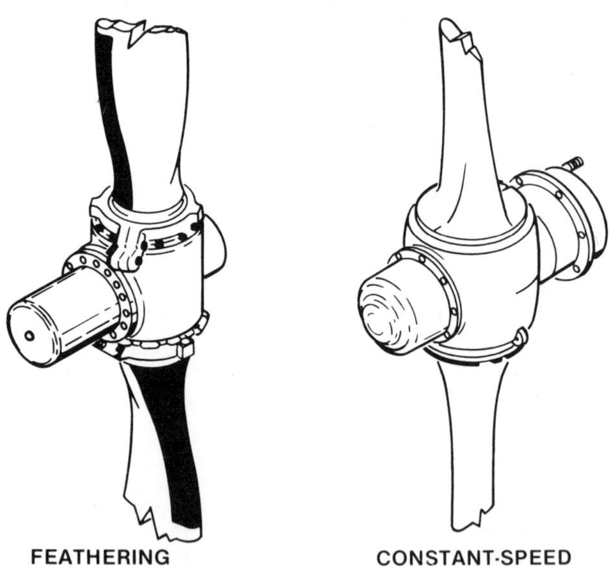

Fig. 4-4 A feathering McCauley propeller compared to a McCauley constant-speed propeller.

Most feathering propellers incorporate a centrifugally operated latch mechanism which locks the blades at a low angle when the engine is at idle on the ground. These latches prevent the blades from freathering when the engine is shut down on the ground. In flight, centrifugal force holds the latches out and they do not interfere with the constant-speed or feathering operation.

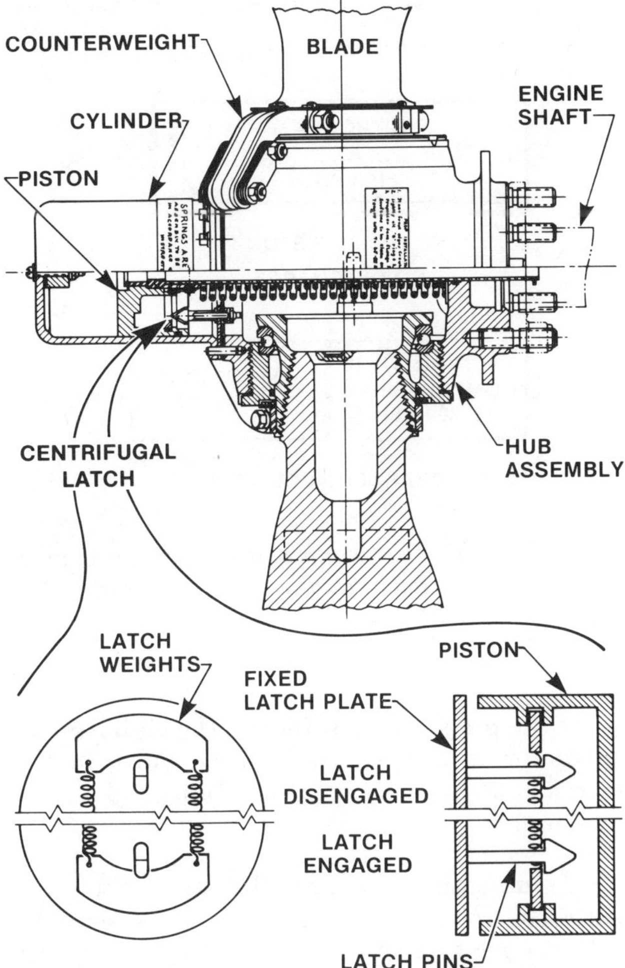

Fig. 4-5 A version of the centrifugal latch mechanism used to prevent a feathering propeller from feathering when the engine is shut down on the ground.

C. Unfeathering Systems

To unfeather a propeller, oil pressure must be supplied to the propeller so that the blades will rotate to a lower angle and start to windmill the engine. There are four methods which may be used to accomplish unfeathering: developing oil pressure by rotating the engine with the engine starter; supplying oil pressure to the propeller from an accumulator; using an electric pump to supply oil pressure to the governor; and, directing oil from another engine propeller system to apply pressure to the feathered propeller.

1. Engine starter

To unfeather the propeller in some aircraft the starter motor for the engine is used to rotate the engine and the governor to develop oil pressure which is directed to the propeller through the underspeed condition of the governor and decreases the propeller blade angle. As the blade angles decrease, the airflow from the aircraft moving through the air will cause the propeller to start to rotate. As the engine and propeller start to rotate, the oil pressure from the governor will increase to the normal governor pressure and the propeller-governor system will return to constant speed operation.

2. Accumulator systems

Many aircraft incorporate an accumulator in the propeller system to be used when unfeathering the propeller. The accumulator is normally located in the engine compartment and is connected to the governor by a flexible or a rigid oil line.

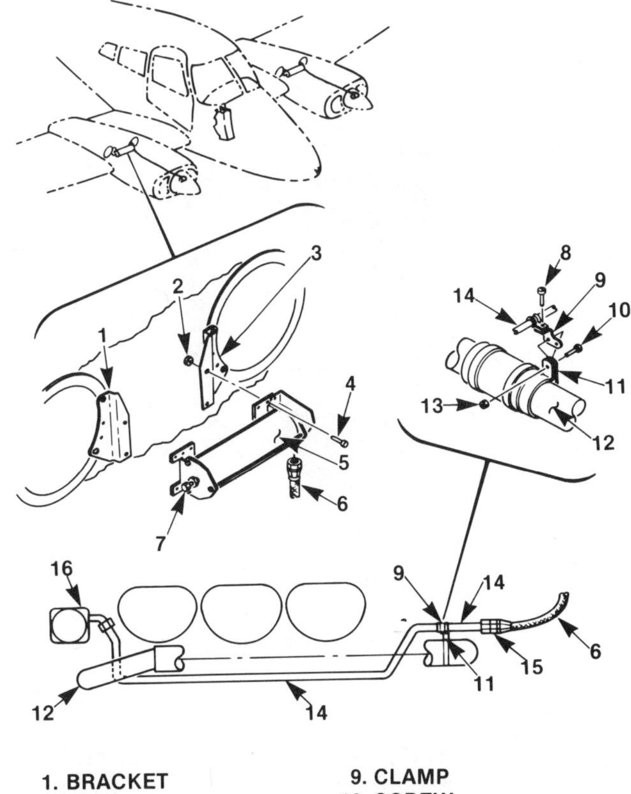

Fig. 4-6 The accumulator system in a Cessna 310.

The accumulator may be of the ball-diaphragm or the piston-cylinder type. In either case, an air charge of about 100 psi will be on one side of the diaphragm or piston when the engine is shut down on the ground and the other side will be connected to the governor oil pressure line for the propeller through the oil line mentioned above.

COURTESY OF WOODWARD GOVERNOR CO.
CYLINDER TYPE

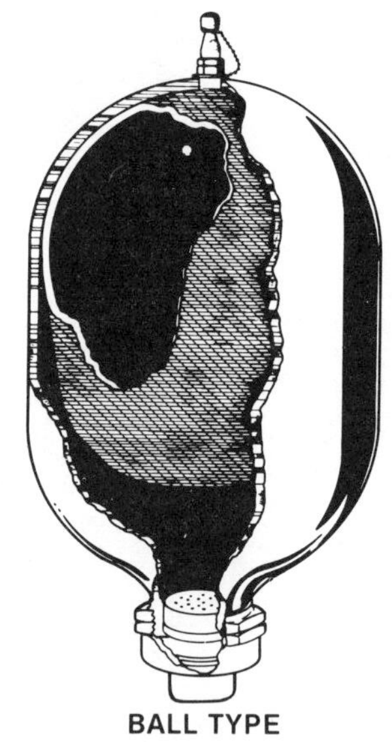
BALL TYPE

Fig. 4-7 Two types of unfeathering accumulators.

During normal constant-speed operation, the governor oil pressure that is applied to the propeller is also directed to the oil side of the accumulator to charge the accumulator with oil at governor oil pressure. Since the governor oil pressure is two to three hundred psi, the oil pressure in the accumulator will compress the air in the accumulator to the same pressure as the oil and thereby store a charge of oil at governor oil pressure.

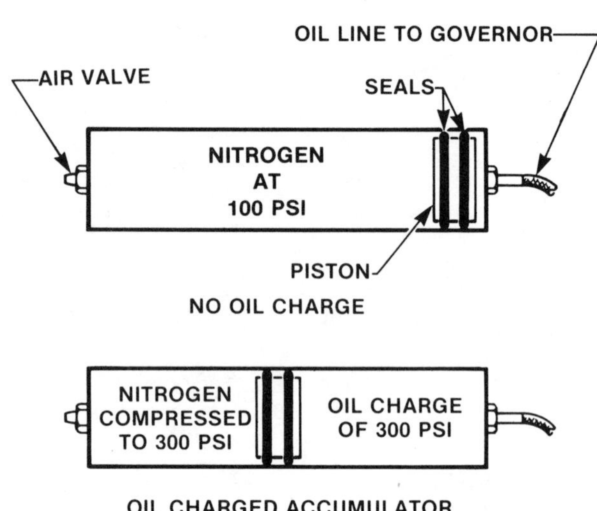

Fig. 4-8 When the accumulator is charged with oil the air in the accumulator is compressed to the same pressure as the oil.

If the propeller is feathered, the oil in the accumulator is trapped by a check valve in the governor or by a control valve attached to the governor.

To unfeather the propeller, the cabin propeller control is moved forward. This sets the governor pilot valve to direct oil to the propeller (governor underspeed condition) and the check valve or control valve is opened and the oil pressure in the accumulator is released into the propeller. The oil pressure moves the propeller to a lower angle, the propeller starts to windmill, and the engine can be restarted.

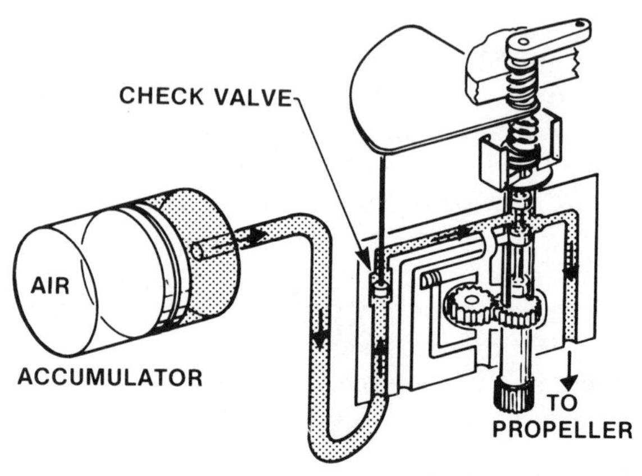

Fig. 4-9 The governor is set to decrease the propeller blade angle when unfeathering with an accumulator.

Once the system returns to constant-speed operation, the accumulator is again charged and is set for the next unfeathering operation.

3. Unfeathering oil pumps

A few aircraft use an electrically operated oil pump to supply oil pressure for unfeathering the propeller. In this unfeathering system an oil line from the engine oil sump is connected to an electric oil pump located in the engine compartment. The output side of the pump is connected to the propeller oil line between the governor and the propeller.

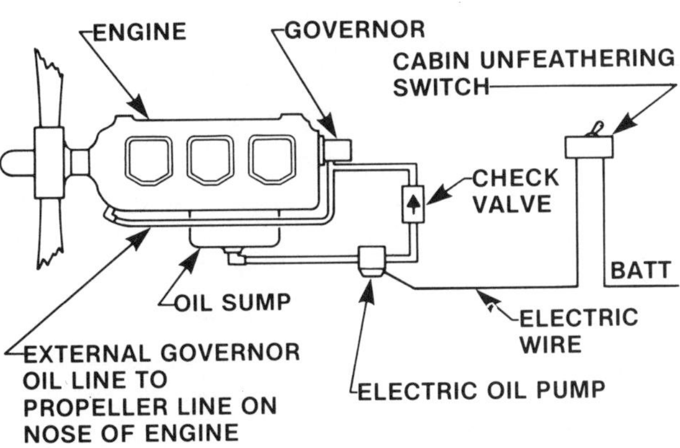

Fig. 4-10 An unfeathering system using an electric oil pump (Reference—STC Piper PA 23-250 SA-6E8CE.)

The oil pump operation is controlled from the cabin by a toggle switch or push button. When the pilot desires to unfeather the propeller, the propeller control lever in the cabin is moved into the constant-speed range and the cabin unfeathering switch is operated until the oil pressure from the electric oil pump causes the blades to move to a low enough angle that the propeller starts to windmill. The switch is then released and the normal engine restart procedure is followed. During this operation, oil pressure from the electric pump is also applied to the governor, but with the governor in the underspeed condition. No significant amount of oil pressure is lost through the governor.

4. Unfeathering crossfeed system

A few aircraft designs use a crossfeed device to tap oil pressure from an operating constant-speed system and direct it to the feathered propeller.

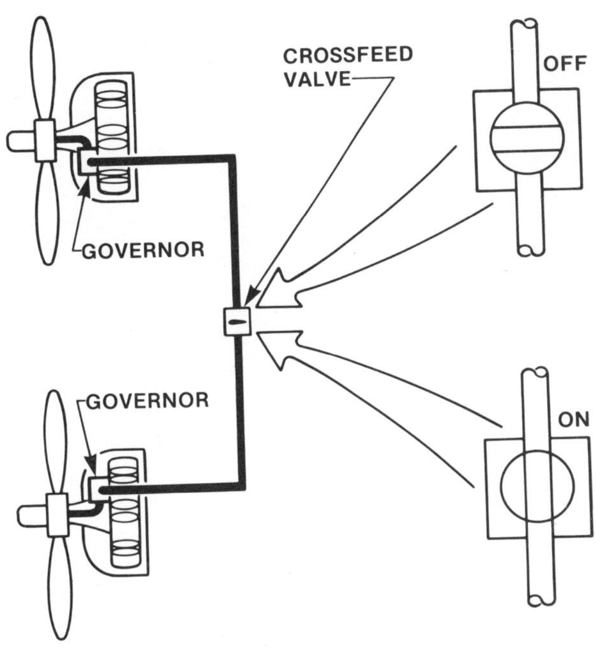

Fig. 4-11 An unfeathering crossfeed system.

To unfeather with a crossfeed system, the propeller control lever for the feathered engine is placed in the full forward position (this may vary for some aircraft) and the crossfeed valve in the cabin is positioned to direct governor oil pressure from the opererating engine to the feathered propeller. This is to unfeather the propeller and cause engine rotation. If the cabin control for the propeller is not positioned properly, the oil from the operating engine may go into the sump of the feathered engine with a loss of oil pressure to the operating engine's propeller and feathering of that propeller!

QUESTIONS:

1. What force is used to decrease the propeller blade angle in most feathering systems?

2. Which governor component is used to override the constant-speed operation and raise the pilot valve?

3. What prevents a feathering propeller from feathering when the engine is shut down on the ground?

4. What is the approximate air pressure in an accumulator when the engine is shut down on the ground?

5. What action is necessary to release the oil charge from the accumulator to unfeather a propeller?

6. Where does the unfeathering oil pump draw oil from to unfeather a propeller?

SECTION V

Governor Installation

The installation and external adjustment of the propeller governor falls within the authority of the aircraft powerplant mechanic.

To overhaul or repair internal components of a governor requires that an overhaul facility approved for propeller governors perform the work.

A. Governor Installation

1. Pre-installation checks

The following types of inspections should be performed before a governor is installed on an engine: (Always consult the appropriate service manual for information about specific installations.)

Check the governor data plate to be sure that the governor is the correct model for the engine and the aircraft. Pay particular attention to any designation referring to the direction of rotation for which the governor is set. At the same time, examine the governor for any external damage and check that all safeties are intact.

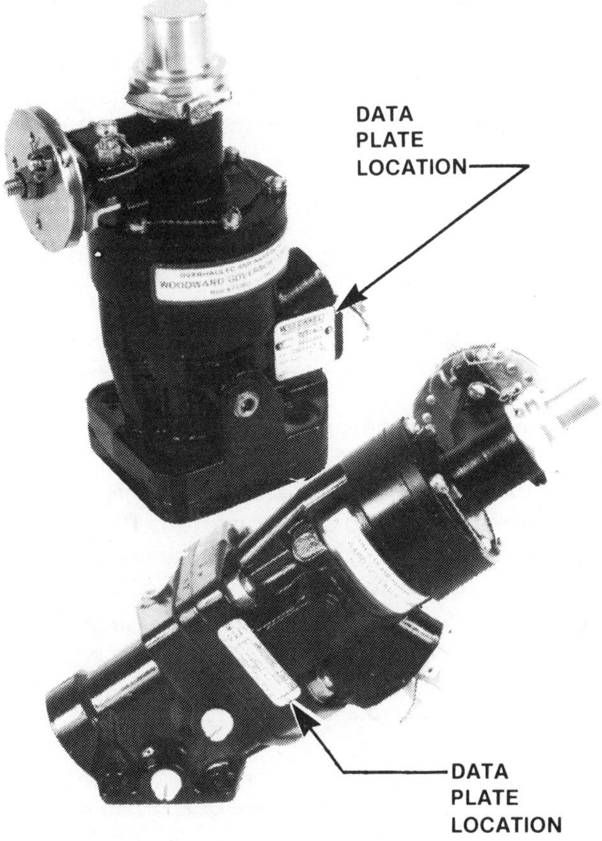

Fig. 5-1 The data plate for a governor should be checked before installing the governor.

Turn the governor drive shaft by hand to check for any grittiness or binding in the gears or drive shaft components. The drive shaft may be hard to rotate due to cold oil or preservative in the governor, but as long as the shaft turns smoothly, the governor may be considered serviceable. The preservative oil used in many governors is compatible with engine oil and does not need to be removed before the governor is installed on the engine. Always verify that the preservative is compatible with the engine oil before installing the governor!

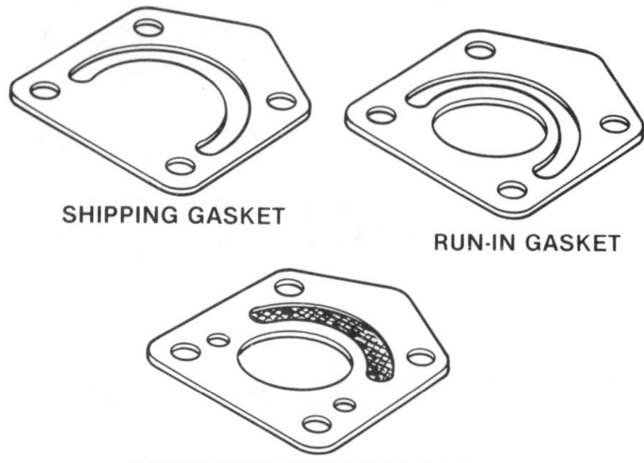

Fig. 5-2 Be sure to use the correct gasket when installing a governor.

The governor mounting surfaces on the engine and the base of the governor should be checked for foreign material, surface damage, and to be sure that the surfaces are free of oil. Check that the governor mounting gasket is the correct part for the particular operational installation and is not a shipping or run-in gasket, and that the gasket filter screen is in good condition. The gasket should be clean, dry, and free of any gasket sealing compound. Some installations require the use of a release agent, which should not be confused with sealing compound.

2. Governor installation

To install the governor, place the governor mounting gasket on the mounting studs with the raised side of the gasket screen away from the engine surface (toward the governor). Place the governor on the engine pad so that the studs pass easily through the mounting holes in the governor. Many governors have the holes spaced so that they will only fit on the engine in one position.

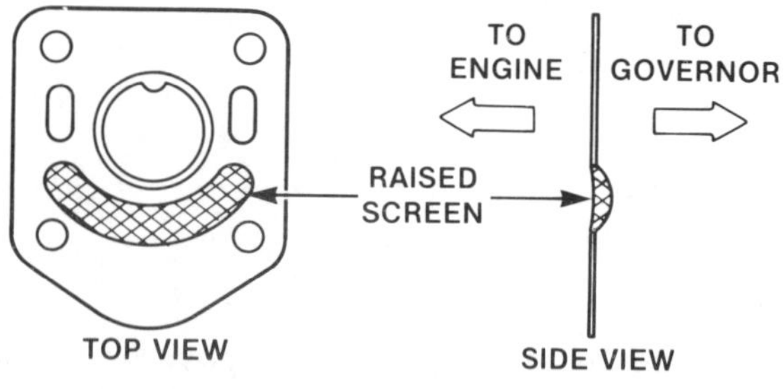

Fig. 5-3 Be sure the mounting gasket is installed with the raised screen toward the governor.

As the governor drive shaft engages the drive gear on the engine, it may be necessary to rotate the engine slightly by moving the propeller to allow the splines on the governor drive shaft to line up with the engine drive gear. The governor should set flush with the engine drive pad.

The proper mounting hardware is now installed, usually a plain washer, lock washer, and plain nut for each stud. The nuts are then torqued to the value specified for the particular installation.

In some installations it is necessary to lift the governor off of the mounting surface to place the hardware on the studs.

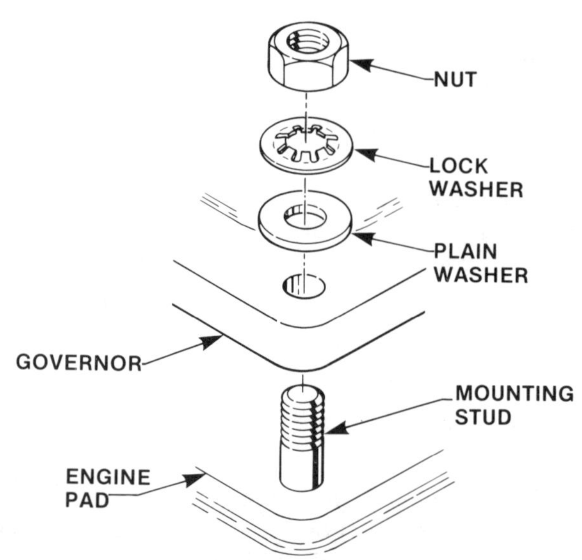

Fig. 5-4 An example of the governor mounting hardware.

The cabin control linkage to the governor is now connected according to the aircraft maintenance manual. On most aircraft this involves using a bolt to connect a push-pull rod end to the governor control arm. The RPM and the cushion of the control linkage is then adjusted.

B. Governor RPM Adjustment

Once the governor is installed, the engine should be operated and allowed to stabilize at operating temperature so that the governor maximum RPM setting can be adjusted and the system can be checked for proper operation.

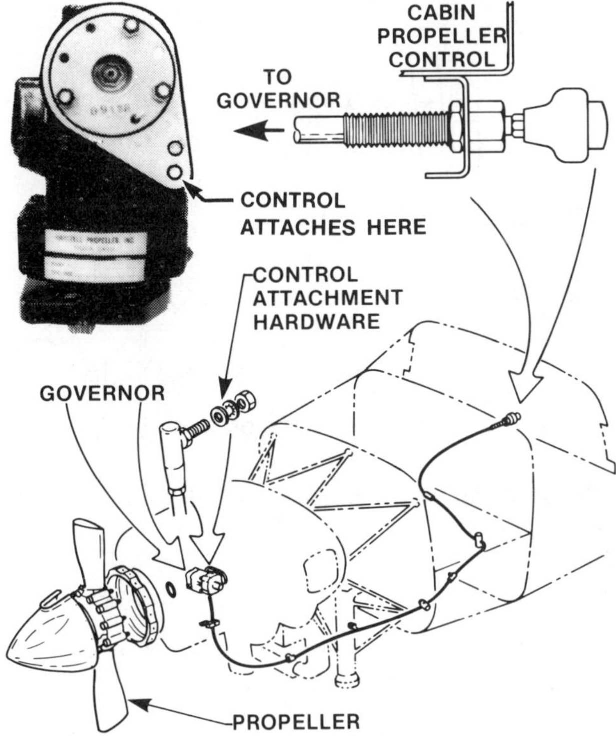

PHOTO COURTESY OF HARTZELL

Fig. 5-5 A typical governor cabin control linkage.

With the throttle at idle, the propeller control in the cabin should be positioned full forward so that the control arm on the governor head is against the high RPM stop. The throttle is then slowly advanced until the RPM no longer increases, indicating that the engine RPM has reached the setting of the governor and the governor is adjusting the propeller blade angle to maintain the RPM at the setting of the governor high RPM stop.

If the RPM is less than the desired high RPM limit, the engine should be shut down and the governor high RPM set screw adjusted to allow a higher RPM (move it away from the control arm so the arm can move further to a higher setting). The governor RPM setting will increase about 25 RPM per revolution of the set screw. (This value varies, depending on the governor and engine model.) Some aircraft cannot achieve maximum RPM on the ground due to the low blade angle setting of the propeller. For these aircraft, maintenance flights will have to be carried out for high RPM adjustment. Always check the aircraft manual for proper adjustment procedures.

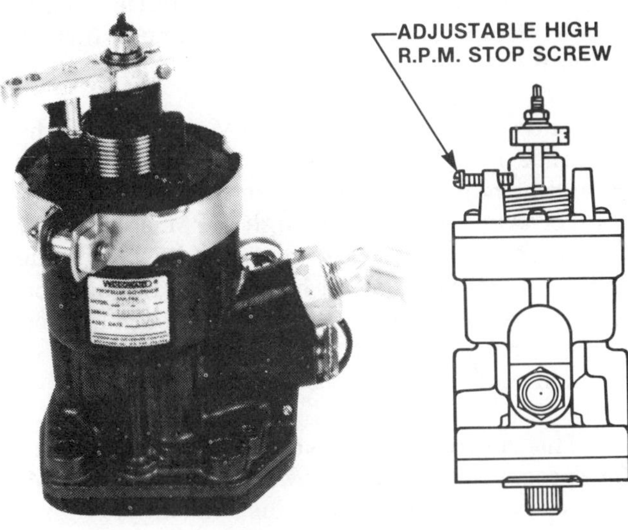

LEVER HEAD TYPE

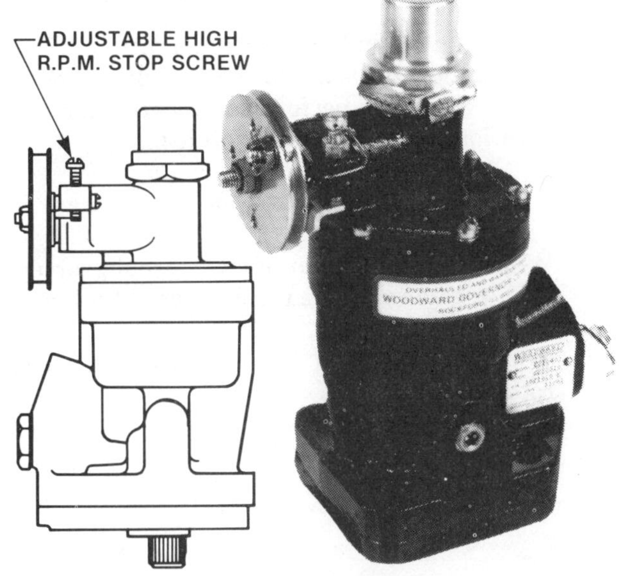

PULLEY HEAD TYPE

PHOTO COURTESY OF WOODWARD GOVERNOR CO.

Fig. 5-6 The set screw sets the high RPM limit of the constant-speed system by limiting the movement of the governor control arm or pulley.

If the RPM begins to exceed the engine maximum RPM during the initial run-up, the advance of the throttle should be stopped before the RPM limit of the engine is exceeded and the cabin propeller control should be moved aft until the RPM starts to decrease. The engine should then be shut down without moving the propeller control. The governor set screw is now adjusted so that it contacts the control arm at the position where the RPM started to decrease.

After either of these adjustments, the engine should be operated again to determine if further adjustment is necesssary.

Once the desired high RPM setting of the governor is achieved, the cabin control linkage should be adjusted in length so that the governor control arm contacts the high RPM stop screw before the cabin control contacts its stop. The cabin control should still have about 1/8-inch forward travel or cushion remaining when the governor control arm is at its stop. The amount of cushion varies, depending on the aircraft.

The control linkage should also be checked for proper movement toward the low RPM/ feather position. If there is insufficient travel aft on a multi-engine aircraft the propeller may not feather or may be slow to feather.

The security and external condition of all system components should be checked and a flight test should be performed to check for proper operation and to check for oil leaks.

QUESTIONS:

1. What is an indication of an internal problem with a governor when it is rotated by hand?
2. What may cause a governor to be hard to rotate by hand?

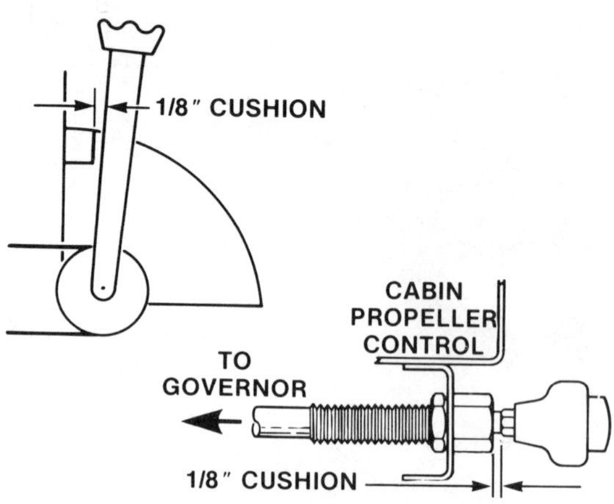

Fig. 5-7 The cabin control should be adjusted for the proper cushion after the governor RPM stop is set.

3. What type of compound may be used on a governor mounting gasket?
4. What is the position of the mounting gasket screen when installing the governor?
5. What is used to limit the travel of the governor control arm in the high RPM direction?
6. What is the approximate cushion desired when adjusting the governor cabin control lever?

SECTION VI

Governor Maintenance

A. Governor System Inspection

When inspecting a governor system, perform an operational check of the system to determine if the governor is functioning properly and to check the high RPM limit of the governor. This should be accomplished before opening up the engine compartment to give the mechanic a better idea of what system components are malfunctioning or are out of adjustment. During the run-up, check the controls for proper response, binding, and cushion.

With the engine shutdown, operate the cabin control through its full range of travel to determine if any problem with the controls occurs only when the engine is operating or only when the engine is shutdown. If there is a difference in the control action, this may be an indication of loose or worn engine mounts or a control resistance problem which is not noticeable or is eliminated by engine vibration.

When removing the cowling and baffling around the governor and control system, check for evidence of wear or abrasion which may lead to, or is the cause of, an operational difficulty.

1. Oil leaks

Before cleaning the area around the governor, look for evidence of oil leaks, especially around the governor fittings, between governor housing parts, at the engine mount, and at any cracks which may have developed in the governor housing.

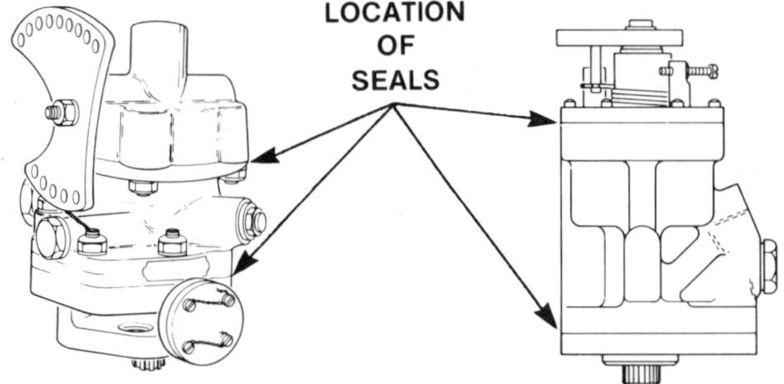

Fig. 6-1 The location of the housing seals on two different styles of governors.

Leaks between housing sections may be caused by loose screws or nuts. Check the torque of all screws and nuts to be sure that they are properly secure. If they are secure and the leak continues, a seal between the housing components may be defective and need replacement. The seals may be replaced by an overhaul facility and, in some cases, by the powerplant mechanic. Consult the appropriate service manual for specific instructions.

Leaks around oil fittings may be caused by loose fittings, broken or cracked fittings, or cracked governor housing components. Tighten or replace the fittings as necessary, or refer the governor to an overhaul facility if the housing is cracked.

Once the governor has been checked for leaks, the area around the governor should be cleaned with an approved solvent and the condition and security of the governor mounting should be checked. Also check to see that all safeties are intact and correctly installed.

2. Check control cable

Check the condition and security of the control cable or push-pull mechanism used to operate the governor. The mechanism should be free of corrosion and should not have any sharp bends or loose support brackets. The security and condition of the fittings which fasten the control linkage to the governor control arm or pulley should also be checked.

Have someone operate the cabin propeller control through its full range of travel and observe the motion of the governor control arm or pulley to be sure that the motion is smooth and allows full travel of the control mechanism.

Any inspection of the governor system should follow the instructions in the aircraft service manual and should also include a check of Service Bulletins, Service Letters, and Airworthiness Directives applicable to the installation.

The recommended overhaul period for governors varies, but is in the range of 1000 hours.

B. System Troubleshooting

To determine the cause of an operational difficulty, the mechanic should have a basic understanding of the relationship between the governor, the propeller, and the engine. He should follow the troubleshooting procedures in the aircraft service manual.

1. Constant-speed systems

The purpose of the governor is to sense changes in system RPM and cause the propeller blade angles to change to correct the RPM variations and maintain a constant-system RPM.

If the power output of the engine is fluctuating, the RPM will vary noticeably even though the governor is operating properly. These variations may be the result of ignition or fuel control malfunctions or the result of a propeller malfunction that will allow the propeller to overshoot the desired angle or be too slow to respond to the governor control input.

Remember, the RPM must change and create an overspeed or underspeed condition in the governor before it will act to change the propeller blade angle. If the governor sensing an underspeed condition causes the blade angles of the propeller to decrease just as the engine power surges due to a fuel flow difficulty, the result will be a surge in RPM caused by the fuel flow and not by a governor system malfunction.

a. Engine oil passages

Presuming that the related engine systems are in proper working condition and the difficulty is related to the governor system, the governor may be the cause, or the engine oil passages associated with the governor may be the cause.

If the governor is suspected as being the defective component, the easiest way to verify that the governor is the cause is by removing the governor and replacing it with a governor known to be in proper working condition. If this does not solve the problem, the problem may be in the oil passage between the governor and the propeller.

In many engines, oil passages between the governor and the propeller can be checked by the use of air pressure. Remove the governor from the engine and apply air pressure to the passage at the governor mounting pad which connects to the propeller. The pressure should be started low and increased to about 100 psi. With a good seal between the governor pad and the air line, the propeller blade angle should change. (This will not work with most feathering propellers.)

If the propeller does not shift and there is a *significant* amount of air coming out of the crankcase breather or the oil filler on a wet sump engine, there may be a crack in the engine case allowing the air to escape from the oil passage or the transfer bearing at the crankshaft may be defective. In either case the engine will have to be disassembled to correct the problem.

If the propeller shifts and there is still a significant amount of air coming out of the crankcase breather or oil filler, there may be a crack in the crankcase, the crankshaft, or the transfer bearing may be worn. A slight amount of air escaping during this pressure check might be considered normal as when performing a differential compression check of a cylinder on the engine.

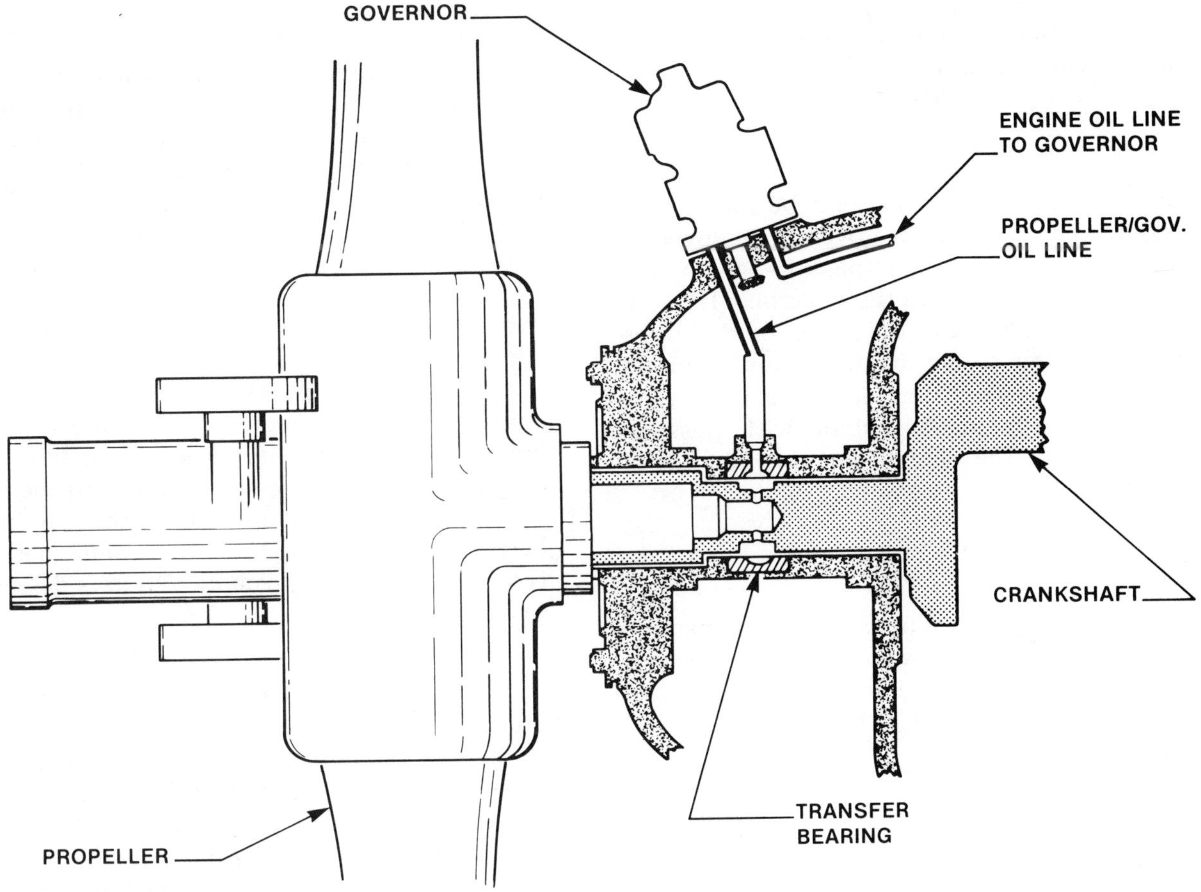

GOVERNOR
ENGINE OIL LINE TO GOVERNOR
PROPELLER/GOV. OIL LINE
CRANKSHAFT
TRANSFER BEARING
PROPELLER

Fig. 6-2 A typical arrangement of the internal engine oil passages used for governor-propeller operation.

If the propeller does not shift and there is no air escaping from the engine, the oil passage between the propeller and the governor may be blocked. The propeller should be removed to determine if air is getting to the propeller. If there is no air coming through or if it is coming through at a very low rate, the oil passage is blocked and should be corrected according to the engine manufacturer's instructions.

On some engines there is an external oil line between the governor and the engine nose section. If there is no air pressure at the propeller from the governor, the external line should be removed and checked for damage or a blockage.

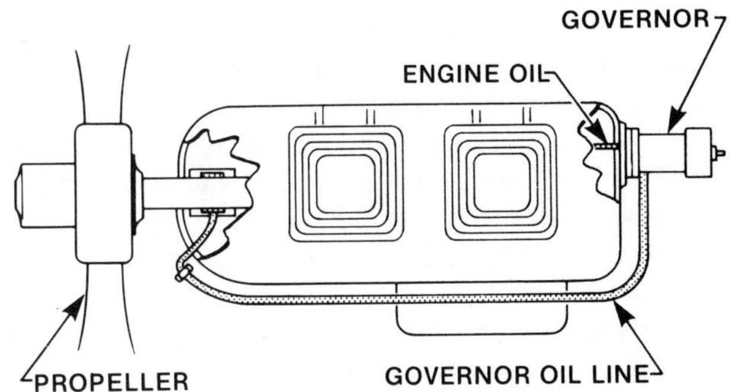

Fig. 6-3 A typical arrangement of external oil passages used for governor-propeller operation.

Whenever applying air pressure for these checks, the air pressure should be regulated so that high pressure is not applied suddenly causing oil or a blockage to suddenly break loose and cause an accident or injury. It is also advisable to hold a rag over the oil passage outlet in case a blockage breaks loose as the pressure is increased.

If the governor change does not correct the problem, and the passage between the governor and the propeller is not blocked, the problem may be with the oil passage delivering oil to the governor.

In some installations this may be checked by rotating the engine with the governor removed to see if oil comes out of the oil passage which supplies engine oil pressure to the governor. If no oil comes out, the passage to the governor should be cleared according to the engine manufacturer's instructions. This condition may be an indication of sludge build-up in the engine with possible engine failure in the near future due to lack of lubrication.

2. Feathering systems

The same basic procedures used for constant-speed systems apply to constant-speed feathering systems. The principle differences are the reasons for improper operation of the feathering and unfeathering mechanisms.

a. Oil pressure

Due to the nature of feathering propellers, if for some reason the governor should lose its oil supply or should stop rotating, oil pressure to the propeller would be lost and the propeller would move to feather, regardless of the power output of the engine or the setting of the cabin controls.

If the propeller feathers for no obvious reason, the oil supply to the propeller was probably lost or the governor drive gear sheared. To determine the cause, the oil passages should be checked for blockages and the operation of the governor should be verified by inspection or replacement with a known good governor.

If the system will not feather, first check to be sure that the cabin control moves the governor control arm to the feather position. If this is operating properly, replace the governor with a known good governor to determine if the governor is defective. If the problem persists, and the propeller is operating properly in the constant-speed range, the blade angle stop for feather on the propeller may be set for too low an angle and needs to be adjusted. To adjust the propeller blade angles, the propeller may have to be referred to a propeller overhaul facility.

b. Cold oil

If a propeller fails to unfeather, there is one condition which cannot be corrected by any means. When the engine is shut down with a low air temperature at the flight altitude, the oil in the engine may congeal. This prevents unfeathering in flight regardless of the type of unfeathering system. However, the system may unfeather without any problem on the ground where it is warmer and the oil is no longer congealed.

If the system relies on the engine starter to supply rotation to operate the engine and governor oil pumps, the electrical system and/or the starter may not be capable of rotating the engine fast enough under conditions of cold oil and low battery power. Take this into account before determining that the govenor system is defective.

c. Accumulator

If an accumulator is used to unfeather the propeller, the proper air charge must be maintained in the accumulator for proper unfeathering action. If the accumulator system air charge is correct and the seals are not leaking between the air and oil chambers, the check valve or control valve at the governor may not be closing properly and is allowing the oil pressure in the accumulator to bleed off. This may be checked by operating the engine on the ground to charge the accumulator and then feathering the propeller.

The charge on the air chamber is checked (it should be about 300 psi when the propeller is feathered) and after 15 minutes, the air pressure is rechecked to see if it has decreased, indicating a loss of oil pressure on the other side of the piston or diaphragm.

The accumulator may be defective and prevent the unfeathering operation by leaking the air charge out through the air valve or by the piston seals or diaphragm being defective and allowing the air charge to pass to the oil side. Both of these conditions can be checked by charging the accumulator and then checking for an air leak with soapy water at the air valve and at the loosened oil line fitting.

d. Oil pump

The unfeathering system which uses an electrically operated oil pump involves basic electrical troubleshooting if the pump will not operate (the motor, the switch, the wires, or the power source may prevent operation). If the electrical components are operating correctly, the oil line to or from the pump may be defective or the pump may be inoperative.

If a crossfeed system fails to unfeather a propeller, the problem is a blocked or defective line between the engines or a defective crossfeed valve. Also, the proper control positioning is critical for unfeathering in some crossfeed systems. Check the procedures being used.

C. Governor Substitution

Many aircraft have more than one governor approved for installation and it may be economical or practical to use a different governor model if the one in use is defective or requires overhaul. The approved governors for an aircraft may be listed in the aircraft maintenance manual, the Aircraft Type Certificate Data Sheet, and the Supplemental Type Certificates for the aircraft.

QUESTIONS:

1. Why should the governor area be inspected before cleaning the area?
2. Why should control movement and cushion be checked with the engine operating and with the engine shut down?
3. What is the easiest way to determine if the governor is the defective component in the constant-speed system?
4. If the engine transfer bearing is worn, what will be the indication when checking the system with air pressure?
5. What could be the result if air pressure is applied suddenly to the engine oil passages?
6. What operational condition will prevent unfeathering of the propeller even when all system components are in good condition?
7. Which system control will prevent an electric pump system from unfeathering if it is defective?
8. Where can a list of optional governors be located?

SECTION VII

Special Governor Designs

This section will discuss governors which have control capabilities beyond the normal constant-speed operation and a system which uses the governors to improve the efficiency of the multi-engine constant-speed systems.

Some propeller governor designs used on multi-engine aircraft provide normal constant-speed operation and, by the addition of external units and some internal design alterations, provide automatic synchronization of the engines. This will set all engines to operate at the same RPM and result in a lower amount of operational vibration for the aircraft.

Other governors are designed to act as safety devices to prevent overspeeding of an engine if the constant-speed governor should fail. These are known as overspeed governors and do not affect the constant-speed system except when a system overspeed occurs.

Some governors are designed to allow the pilot to directly control the propeller blade angle during ground operation and allow conventional constant-speed operation to occur during flight operations. These types of governors are found on some light turboprop installations.

A. Synchronization System Governors

A synchronization system is used on many multi-engine aircraft to keep all engines operating at the same RPM. This reduces the noise and vibration created by engines operating at different speeds and reduces the stress imposed on the pilot by the continuous "beat" of the engines.

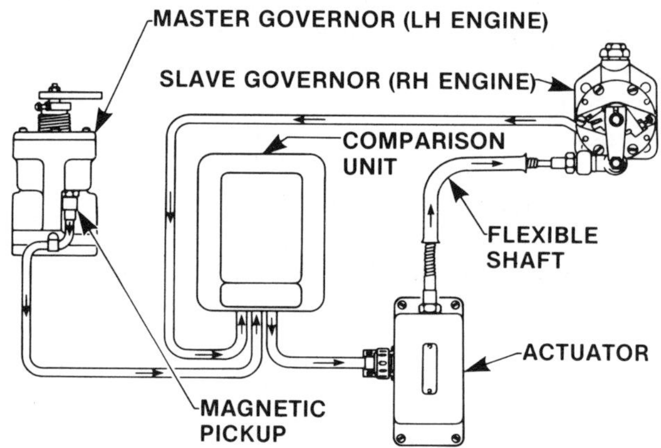

Fig. 7-1 Schematic diagram of a synchronization system. Note direction of signal path during operation.

1. Basic system components

A synchronization system uses one of the aircraft's engines (called the master engine) to set a reference RPM and adjusts the RPM of the other engines (called the slave engines) to the RPM of the master engine.

The system requires the use of three basic units for operation: a unit to sense the RPM of each engine, usually the governor; an electronic unit to compare the RPM of each slave engine to the RPM of the master engine and generate a correcting signal to be sent to the actuator unit; an actuator unit for each slave engine which receives the correcting signal from the comparison unit and adjusts the RPM setting of the engine to that of the master engine.

a. The sensing unit

The propeller governor is often used as the sensing unit for the synchronization system on light aircraft.

A common method used to sense the RPM of the engine through the governor is to use a flyweight plate or cup that is notched evenly on the outer edge. Other designs have raised segments machined or attached to the base of the flyweight cup or plate at the outer edge. As the flyweight unit rotates, the segments pass close to a magnetic pickup mounted in the governor. A pulse from the magnetic pickup is generated as each segment passes the pickup and the frequency at which these pulses are generated is proportional to the RPM of the engine.

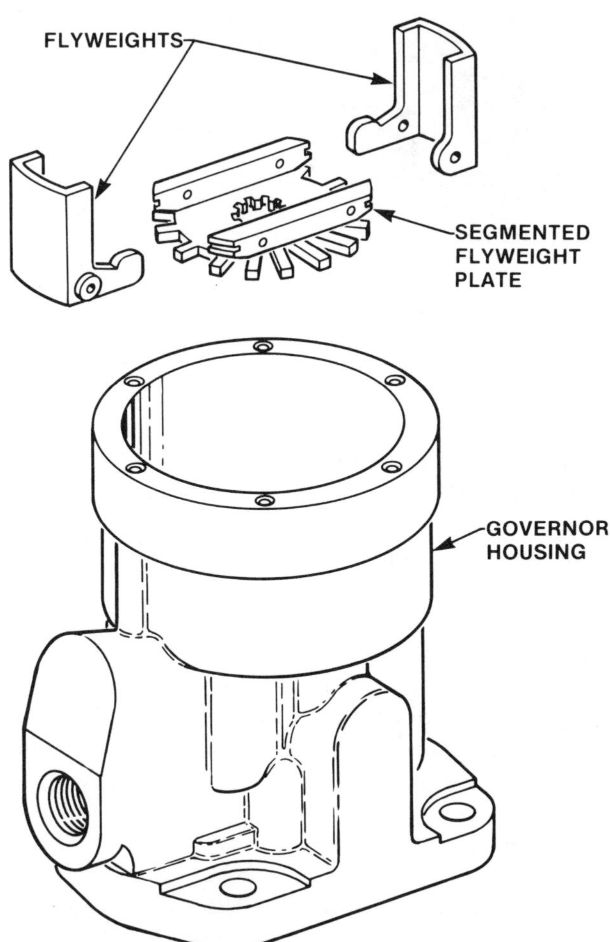

Fig. 7-2 A segmented flyweight plate used on a synchronizing governor to sense system RPM.

A synchronizing governor can often be identified by the wires from the magnetic pickup coming out of the governor.

b. The comparison unit

The comparison unit receives the signal from the sensing units of all of the engines and compares the RPM of the master engine to the RPM of the slave engines.

Through the electronic circuits in the comparison unit, a signal is generated to be sent to the actuator unit on each slave engine. The actuator unit adjusts the governor setting of the slave engine so that it operates at the same RPM as the master engine. The internal circuitry of the comparison unit is beyond the scope of this text. The comparison unit is usually located in the aircraft cabin.

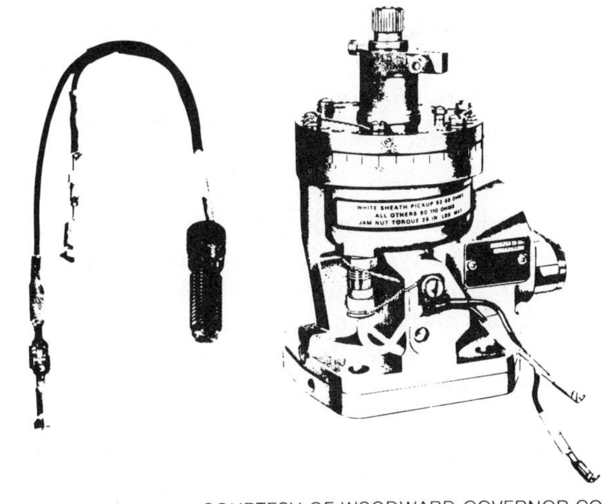

COURTESY OF WOODWARD GOVERNOR CO.

Fig. 7-3 A magnetic pickup is located on the side of the governor and sends a pulsed signal to the comparison unit.

c. The actuator unit

The actuator is usually located in the engine compartment and is used to make small adjustments to the propeller governor control arm setting to set the slave engine to the same RPM as the master engine.

The actuator receives the correcting signal from the comparison unit and rotates a flexible driveshaft which is attached to a special governor

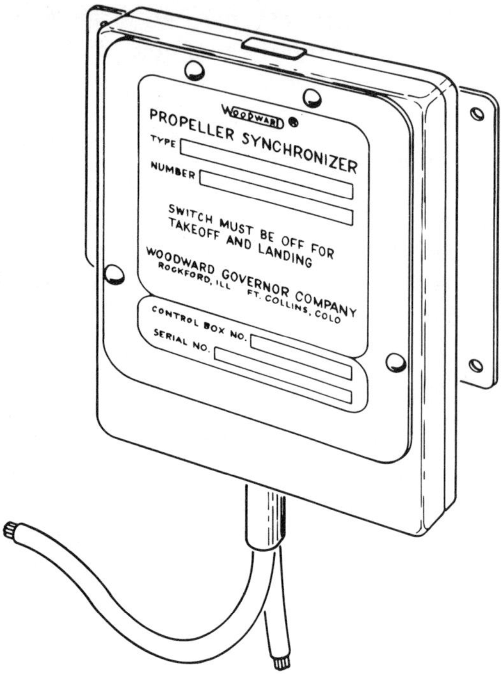

Fig. 7-4 The synchronizer comparison unit is an electronic computer device. Repair of internal components of the unit is usually referred to an overhaul facility.

COURTESY OF WOODWARD GOVERNOR CO.

Fig. 7-5 The actuator is shown with the cover installed and removed. The cover must be removed for lubrication and adjustment.

control rod end. Depending on the direction of rotation of the drive shaft, the rod end will extend or retract from its neutral position to adjust the position of the governor control arm so that the governor will be set for the same RPM as the master engine.

There is an actuator unit for each slave engine, but an actuator unit is not used with the master engine as this is the reference engine for the synchronization system.

d. The cabin control

The synchronization system cabin control consists of a toggle switch to turn the system on and off. Many systems include an indicator light next to the toggle switch which illuminates when the system is turned on.

2. System operation

The system is activated by turning on the cabin switch. If the slave engines are within about 100 RPM of the setting of the master engine, the system will adjust the slave engines to operate at the same RPM as the master engine. (The slave engines have to be within about 100 RPM of the RPM of the master engine in some systems for the synchronization system to synchronize, this value will vary with different systems.)

Before turning on the system, the aircraft is established in a stabilized flight condition, such as a climb. The pilot synchronizes the engines as close as possible and then turns on the synchronization system with the cabin toggle switch.

a. Comparison unit

The signals generated by the governors are sent to the comparison unit where the RPM of each slave engine is compared to the RPM of the master engine. The comparison unit then sends an electrical signal to the actuator unit for each slave engine and commands it to increase, decrease, or hold the RPM setting of the slave

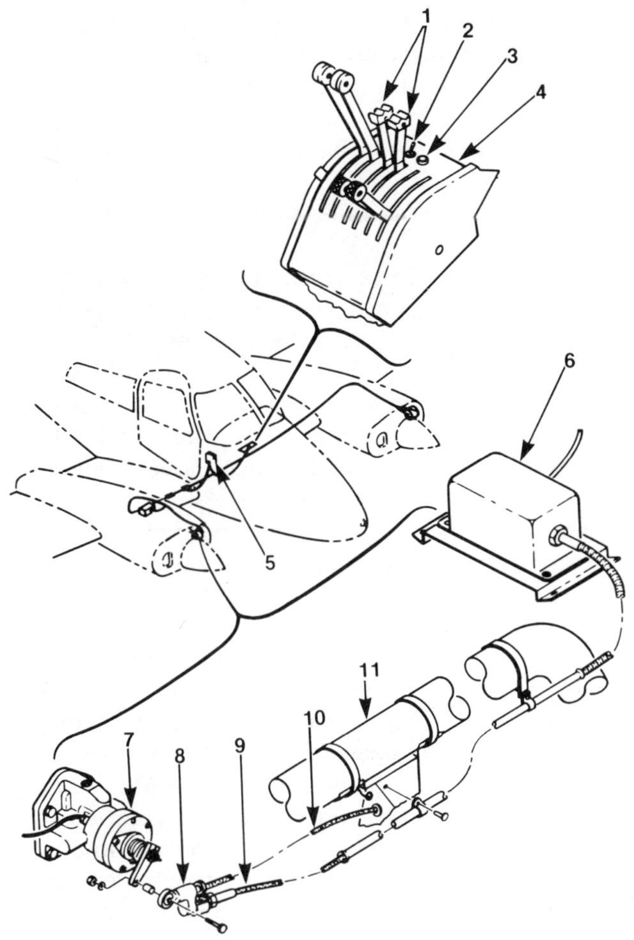

1. PROPELLER PITCH LEVERS
2. SYNCHRONIZER SWITCH
3. INDICATOR LIGHT
4. UPPER PEDESTAL PANEL
5. COMPARISON UNIT
6. ACTUATOR MOTOR
7. GOVERNOR
8. TRIMMER ASSEMBLY
9. FLEXIBLE SHAFT
10. PROPELLER CONTROL CABLE
11. INTAKE MANIFOLD OF ENGINE

Fig. 7-6 System installation on a light twin-engined aircraft.

governor. Each actuator unit receives an adjusting signal which is appropriate for that slave engine and all slave engines will not always receive the same adjustment. One may have to increase and another may have to decrease RPM. The appropriate adjustment for each governor is determined by the computer circuitry in the comparison unit.

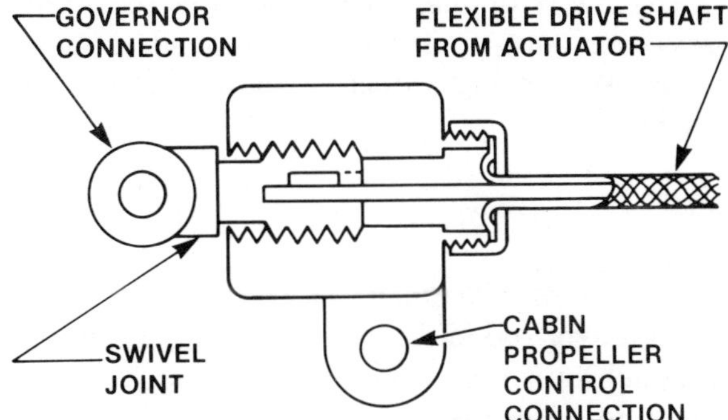

FULL DECREASE RPM POSITION

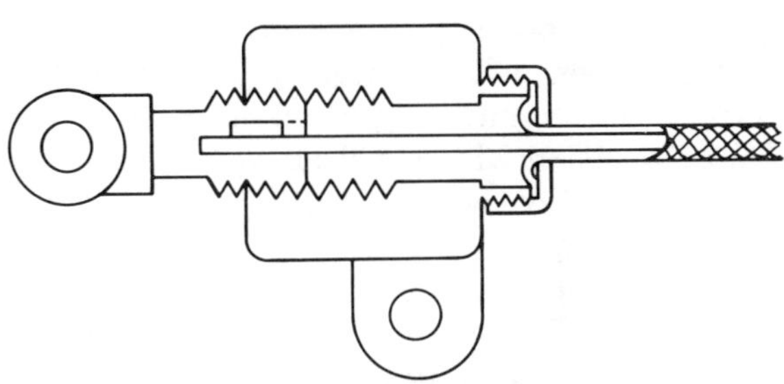

FULL INCREASE RPM POSITION

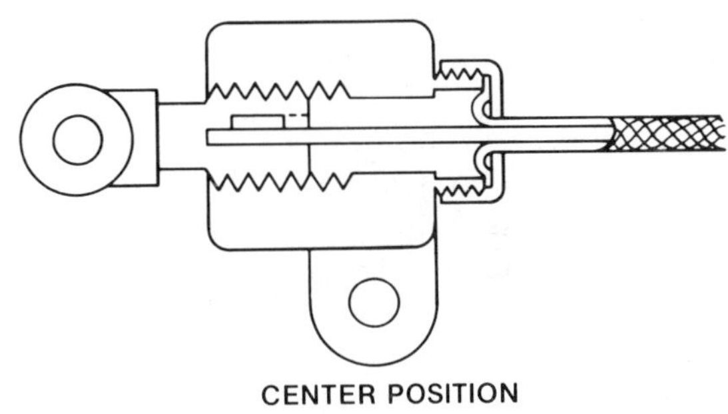

CENTER POSITION

Fig. 7-7 Governor control rod end used with a synchronization system shown in the mid-range (neutral) and extreme fore and aft positions.

b. Too low—too high

If the RPM of a slave engine is too low, the actuator will receive a signal which will cause it to rotate a flexible drive shaft to extend the special governor control rod end which moves the governor control arm forward to increase the governor RPM setting. The forward movement of the control arm will continue until the RPM signal generated by the slave engine is the same as the RPM signal from the master engine.

If the RPM of the slave engine is too high, the governor control arm will be moved aft to a lower RPM setting until synchronization is established.

c. Actuator unit

The limit of movement of the actuator unit and the control rod end determine the maximum out-of-synchronization that can be corrected by the system. Because of this limitation, the engines must be synchronized or close to synchronization before the system is turned on.

When the system is turned off with the toggle switch, power will continue to be supplied to the actuator units until they have returned to the mid-range position which will allow full range of travel above and below the center position.

If the system is turned on when the RPM of the slave engine is sufficiently different from the speed of the master engine to prevent system synchronization, the actuator will drive to its limit to try to achieve synchronization.

d. RPM engine adjustment

The RPM of all engines can be adjusted while the synchronization system is engaged by moving all of the propeller control levers in the cabin in the normal manner. The system should keep all engines synchronized during the RPM change.

The synchronization system is usually not used during takeoff and landing. If the master engine should fail during a takeoff, the slave engines would try to follow a master engine RPM loss and would further reduce the aircraft power available. The slave engines would lose about 100 RPM maximum due to the range limits of the actuator system. Also, if the master engine should fail during a go-around from a landing approach, the same situation would result.

3. Inspection and maintenance

The inspection of the synchronization system involves a visual inspection of the units of the system for security of mounting, condition of wiring, and mechanical condition of components such as the rod ends and the actuator unit. An operational check of the system is performed to determine proper system action.

The following is a general procedure for an operation check of a synchronization system. Refer to the appropriate service manual for an operational check of a particular system.

The engines must be operating at a flight RPM for an operational check. This may be performed on the ground or in flight.

Synchronize the engines with the propeller controls and then turn on the synchronization system — the engines should stay in synchronization. Move the propeller control lever of each slave engine forward until the engines go out of synchronization (as indicated by the "beat" of the engines) and note the distance moved—usually 1/2 to 3/4 control knob width. Repeat the operation aft of the center position. The distance forward and aft of the center position should be about the same.

Next, turn off the system, synchronize the propellers with the cabin controls and turn the system back on. Move the master engine propeller control lever forward until the engines go out of synchronization and note the RPM change and the position of the control knob in relation to the position of the slave engine control knobs. Repeat the procedure moving the master engine propeller control lever aft. The position of the knob fore and aft of the center position should be about the same and the increase and decrease in system RPM from the center RPM originally set should be about the same.

Now, turn the system off and place the engines out of synchronization about 25 to 50 RPM. Turn the system back on and the engines should synchronize. This completes an operational check of the system.

Maintenance of system components is minimal. Other than routine lubrication of some components such as the actuator, it usually involves identification and replacement of defective components. Defective units are usually sent to a repair facility.

B. Overspeed Governors

Some propeller control systems incorporate a safety device called an overspeed governor which is used to increase the propeller blade angle if an overspeed (above system RPM red line) of about 4% occurs. This type of governor is incorporated in some light turboprop installations. The governor is not controllable by the pilot, except for a test function, and operation is automatic.

1. Governor components

The governor contains the same basic propeller oil control components as a constant-speed governor—a speeder spring, flyweights, and a pilot valve. The governor does not contain an oil pump as its purpose is to release oil from the propeller and cause an increase in propeller blade angle to prevent excessive overspeeding. A counterweighted propeller is usually associated with this type of governor. The propeller uses oil pressure to decrease blade angle and the force of springs and counterweights to increase the blade angle.

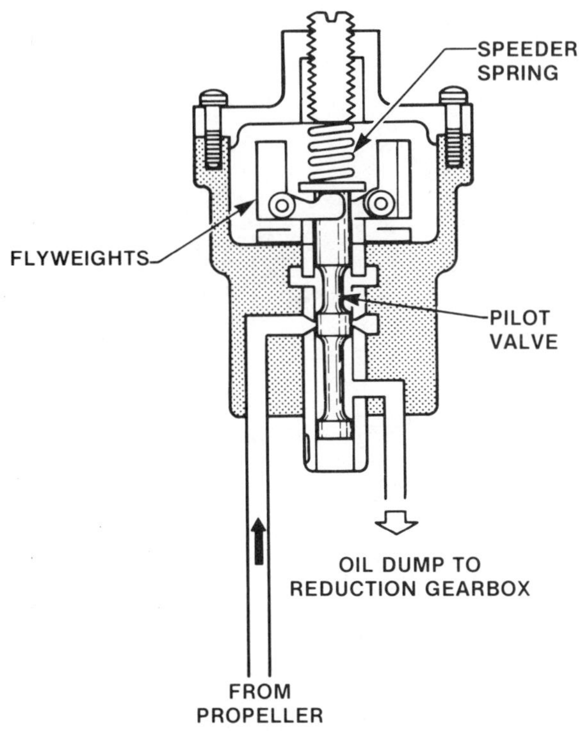

Fig. 7-8 A basic overspeed governor.

The compression of the speeder spring is adjusted by a speed adjustment screw on the top of the governor. This adjustment is normally performed by an overhaul facility as a test stand is required for proper adjustment.

Depending on the particular governor model, the design may incorporate a magnetic pickup used in a synchronization system, a reset solenoid to allow a ground test of the governor at a reduced RPM setting, an unfeathering adapter plate to allow unfeathering of a propeller, or an autofeather solenoid for an automatic feathering system.

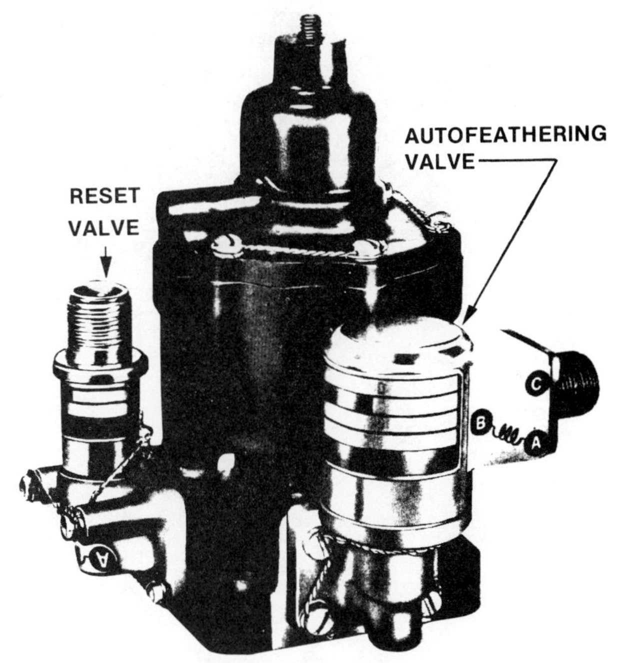

COURTESY OF WOODWARD GOVERNOR CO.

Fig. 7-9 An overspeed governor with a reset valve and an autofeathering valve installed.

The reset solenoid, also called an electric speed reset valve, is energized for a ground test during engine run-up. This solenoid allows oil to flow to the reset piston at the speed adjustment screw and reduces the compression of the speeder spring. This sets the overspeed governor for an RPM of about 90% of the rated RPM of the constant-speed system. When the solenoid is deenergized, the oil pressure on the reset piston is released and the overspeed governor returns to normal operation.

If an unfeathering adapter plate is included in the governor, it will allow oil from an unfeathering pump to enter the propeller control system and apply oil pressure to the propeller for unfeathering. The unfeathering adapter plate incorporates a check valve to prevent a loss of oil pressure through the unfeathering pump during normal operation.

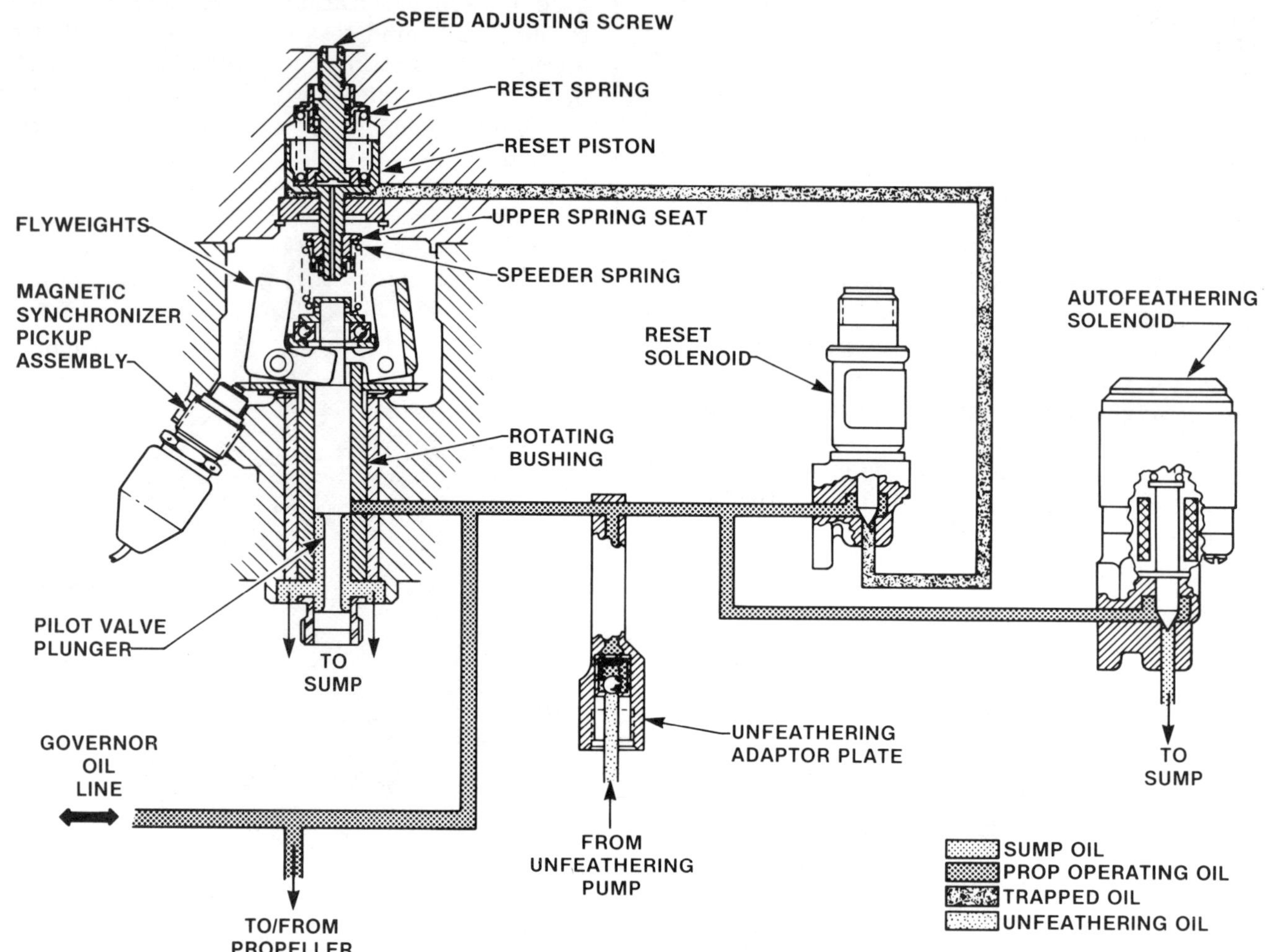

Fig. 7-10 Overspeed governor schematic showing various optional equipment and the associated oil lines.

An autofeathering solenoid is incorporated in some overspeed governors to allow the release of all pressure in the propeller and all of the output volume from the constant-speed governor so that the propeller will move to feather. The solenoid operation is controlled by the engine negative torque sensing system and is energized to allow the release of oil on command of the negative torque system.

2. *Operation*

The operation of the overspeed governor is automatic and is not controlled by the pilot. If the engine or propeller gear reduction unit which is driving the governor exceeds the setting of the overspeed governor, the flyweights in the governor will overcome the force of the speeder spring and move outward. This will raise the pilot valve and release the oil pressure in the line between the propeller governor and the propeller. This causes the propeller blade angle to increase by the force of the propeller springs and counterweights.

The overspeed governor will remain in an underspeed conditon below the 104% system RPM value. This keeps the pilot valve lowered and closes the drain line and allows the propeller control system to function normally. (104% is a representative value. It may be different for specific installations.)

During engine run-up prior to a flight or during a maintenance ground check, the operation of the overspeed governor can be checked by the use

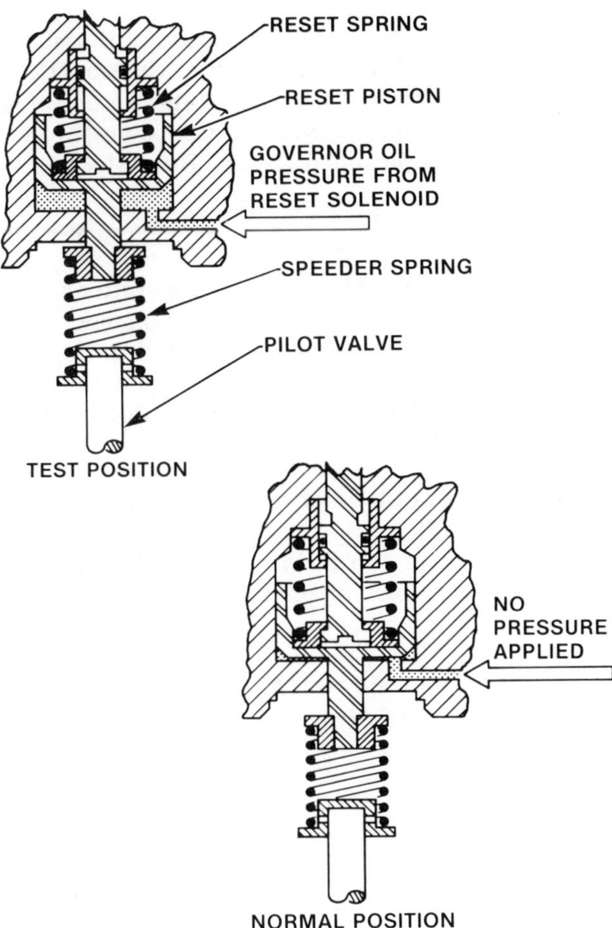

Fig. 7-11 Overspeed governor speeder spring area with the reset valve in the normal and the test position.

of the electric speed reset valve. This opens a passage to the reset piston above the speeder spring and raises it to increase the compression on the speeder spring. This results in the governor being set for about 90% of the system rated RPM. As the RPM of the system is increased to the 90% value, the pilot valve in the overspeed governor will rise and oil in the propeller line will be released as necessary to prevent an RPM above 90%.

3. Installation and inspection

The installation of the overspeed governor is basically the same as for any governor. However, depending on the other capabilities associated with the particualr model governor being installed, an external line may have to be connected for the unfeathering operation, and electrical connections may have to be made for the speed reset valve, the synchronizer pickup, and the autofeather system solenoid.

Inspection of the governor involves checking the system for operation with the reset valve and inspecting the external safeties and condition of the surfaces of the governor for cracks and damage. The security of the installation and condition of the electrical connections and oil lines should also be checked.

The repair and adjustment of an overspeed governor should be referred to an overhaul facility.

C. Governors With Manual Control

Some light aircraft are equipped with governors which function in a normal constant-speed manner in flight (known as Alpha Mode), but allow the pilot to have direct control of the propeller blade angle during ground operation (known as Beta Mode). These governors are usually found on light aircraft having reversing propeller systems. The constant-speed control of the governor is adjusted by the cabin propeller control lever and the ground control of the propeller blade angle is through a throttle linkage to the governor.

For this discussion a type of governor similar to that used on some versions of the Pratt and Whitney PT6 turboprop system will be used to illustrate the basic principles of operation.

1. Basic design

The governor contains the basic components found in all constant-speed control governors: an oil pump, relief valve, pilot valve, speeder spring, flyweights, and governor control lever. In addition, the governor contains a Beta control valve which is used to direct oil to and from the propeller during ground operations to control the propeller blade angle.

During Beta Mode (ground) operations, the Beta control valve is positioned by the movement of the cabin throttle control (termed the power lever in turbine engine systems).

Constant-speed operation does not occur during Beta Mode operations.

The system requires the use of a slip ring on the rear of the propeller which is attached to the

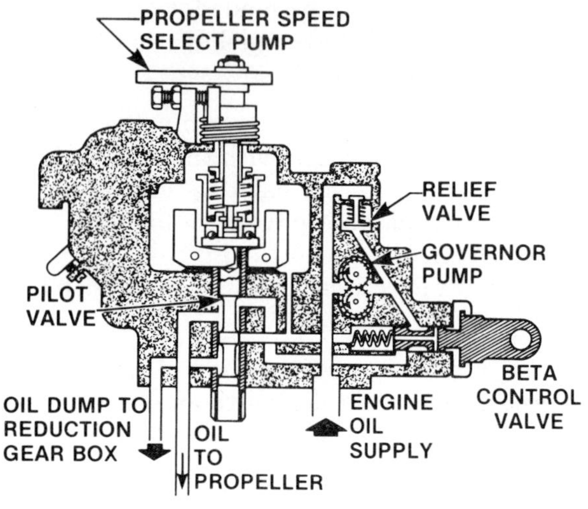

Fig. 7-12 Cross section of a governor similar to that used with a PT6 installation. Note the basic constant-speed components and the addition of a Beta control valve.

propeller piston so that the position of the piston can be compared to the position of the throttle to determine the correct position of the Beta control valve. This comparison of position gives proportional blade angle control in the Beta Mode.

2. *Operation*

During Alpha Mode (flight) operations, the governor functions in a conventional constant-speed manner. Oil pressure is directed to the propeller to decrease the blade angle and released from the propeller to increase the propeller blade angle.

In the Beta Mode of operation, movement of the throttle will cause the blade angle to change by the action of the Beta control valve and the slip ring on the rear of the propeller.

If the pilot wishes to move the aircraft forward in the Beta Mode, the throttle is moved forward. This causes the Beta control valve to be

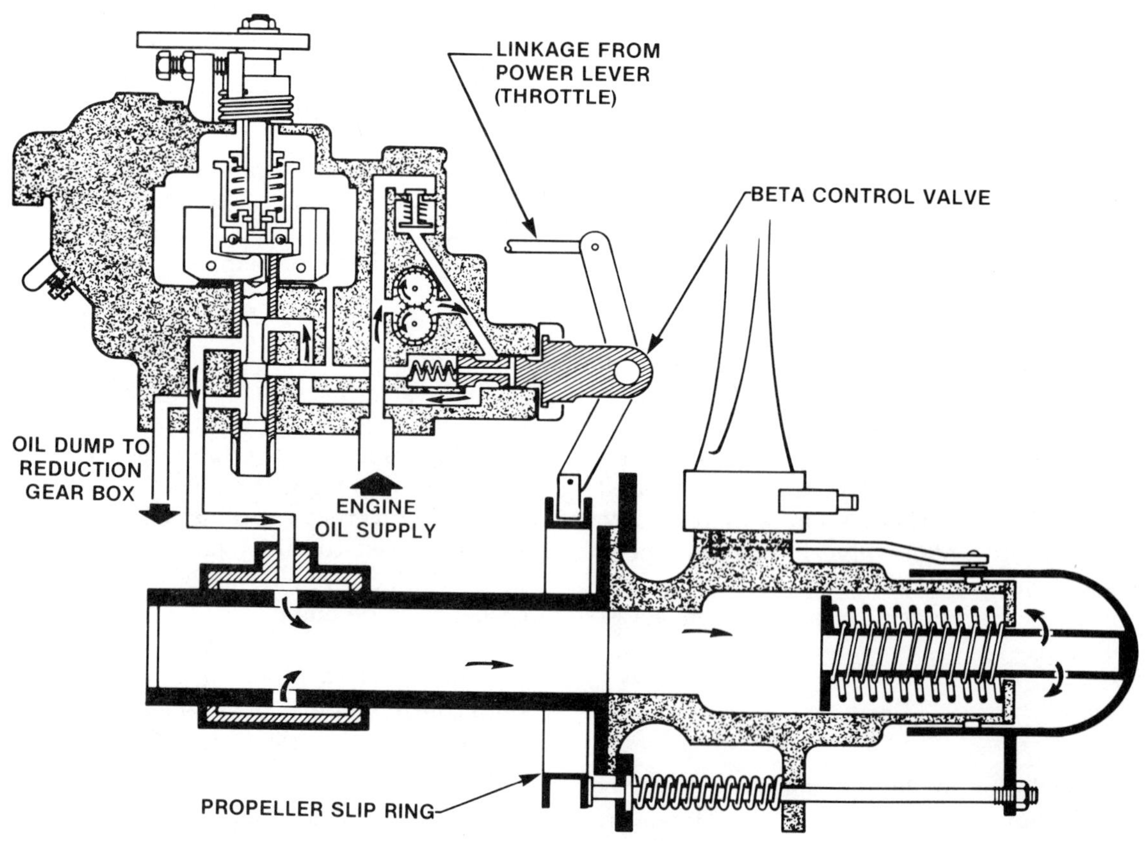

Fig. 7-13 Typical position of the governor in relation to the propeller and the interconnection between the propeller slip ring and the Beta control valve.

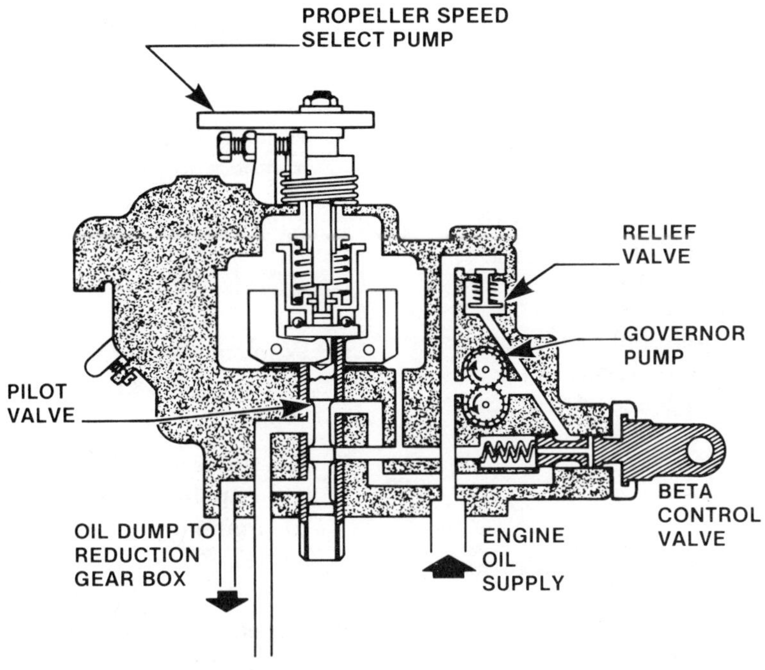

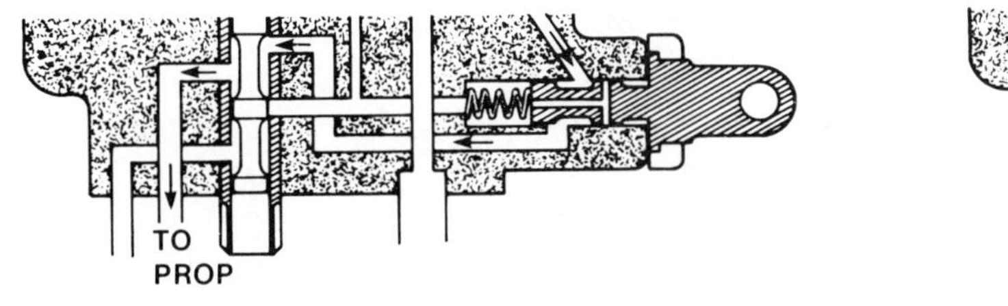

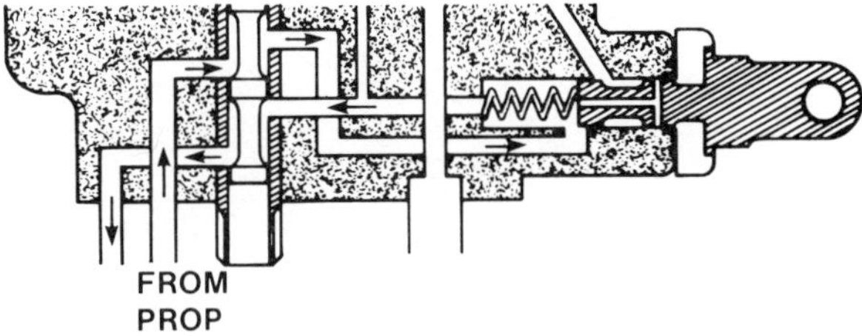

Fig. 7-14 The positions of the Beta control valve in the propeller governor to increase, hold, and decrease the propeller blade angle.

moved forward, out of the governor body, and allows oil to flow from the propeller, past the Beta control valve, to the engine oil sump.

As the oil flows out of the propeller piston, the piston moves rearward which moves the slip ring rearward and increases the propeller blade angle by the force of springs and counterweights. This returns the Beta control valve to the center position through the linkage between the throttle linkage, the Beta control valve and the propeller slip ring.

If the throttle is moved forward a large amount, the propeller blade angle and slip ring travel must be proportionally large to return the Beta control valve to the neutral positon. If the throttle is moved a small distance the blade angle will only change a small amount until the Beta control valve is returned to the neutral position.

If the pilot wishes to slow the aircraft or reverse the blade angle for braking, he moves the throttle aft. This moves the Beta control valve into the body of the governor and directs oil from the governor oil pump to the propeller oil line in the governor. As oil flows into the propeller the piston fills with oil and moves forward decreasing the blade angle, moving the slip ring forward, and returning the Beta control valve to the neutral position.

3. Installation and maintenance

The installation of the governor is basically the same as for any constant-speed governor with the exception that the throttle lever linkage and the linkage between the propeller slip ring and the Beta control valve must be rigged.

The maintenance for the system varies from the basic constant-speed system in that several rigging adjustments must be inspected and adjusted and the linkage between the throttle, Beta control valve, and propeller slip ring must be inspected for wear and damage.

The propeller slip ring rotates with the propeller and the connecting linkage to the Beta control valve is stationary with a carbon block on the end of the linkage riding in the slip ring groove. This block is subject to wear and should be checked at regular intervals in accordance with the aircraft service manual.

QUESTIONS:

1. What is the purpose of a synchronization system?
2. What are the three basic units of a synchronization system?
3. What generates an RPM rate signal in a synchronizing governor?
4. What is the term for the reference engine in a synchronization system?
5. What is the purpose of an overspeed governor?
6. What is the purpose of the reset solenoid of an overspeed governor?
7. If a governor has a manual blade angle control capability, which cabin control is used to manually change the propeller blade angle?
8. Which governor component is used to change the propeller blade angle manually in the Beta Mode?

GLOSSARY

This glossary is provided to serve as a ready reference for the words with which you may not be familiar. These definitions may differ from those of standard dictionaries, but are in keeping with aircraft system terminology.

accumulator An air-charged cylinder which is used to store a charge of oil to unfeather a propeller.

actuator unit The component of a synchronizer system which receives an adjustment command from the comparison unit and converts it into a mechanical movement to move the speed adjustment lever on the governor.

Alpha Mode A term usually related to reversing systems indicating an operation associated with takeoff and flight.

autofeather system A system which senses the power output of an engine and will cause the propeller to be feathered automatically, without pilot action, if engine power is lost.

base The bottom portion of a governor which includes the governor surface used for mounting the governor on the engine.

Beta control valve The oil flow control valve used in some governors to allow the pilot to have direct control of the propeller blade angle when operating in the Beta Mode.

Beta Mode A term usually related to reversing systems indicating an operation associated with ground operations and landing reverse thrust.

blade angle The angle between a propeller blade station and the plane of rotation of the propeller.

body The center portion of the governor which contains the governor oil pump, relief valve, and oil passages to direct oil flow to and from the pilot valve.

check valve A valve usually composed of a ball and spring. The spring is used to seat the ball and block a passage. The ball may be unseated by oil pressure, as in a pressure relief valve, or by a push rod, as in a governor used with an accumulator system.

comparison unit The component of a synchronizer system which receives an RPM signal from the aircraft engine RPM sensors, compares them, and generates a correcting signal to be sent to the slave engine actuator units.

constant-speed system A system which will vary the propeller blade angle as necessary to maintain a set or selected system RPM.

crossfeed To direct a fluid from one system to a second similar system as when directing governor oil pressure from an operating propeller system to a feathered propeller system to unfeather the propeller.

drive shaft The central shaft of a governor which attaches to an engine drive gear and rotates the governor oil pump to generate oil pressure and rotates the governor flyweights to sense system RPM.

feather When the propeller blades are rotated to an approximate 90 degree blade angle to reduce the drag of the propeller when an engine fails.

fixed force A force acting on a propeller at all times during propeller rotation which is trying to change the propeller blade angle; i.e., springs, centrifugal force acting on counterweights.

flyweights The weights in the head of the governor which are used to sense engine speed. The centrifugal force or rotation causes the flyweights to move outward or inward in conjunction with the force of the speeder spring and position the pilot valve in the governor to direct oil to or from the propeller.

governor The component of the constant-speed system which senses system RPM and directs oil pressure to or from the propeller to change the blade angle to maintain the RPM selected by the pilot.

head The top portion of the governor which usually contains the speeder rack, and RPM adjustable stops.

housing The castings which form the containers for the components of the head, body, and base of the governor.

lift rod The rod located above the pilot valve in many feathering governors which is used to lift the pilot valve and release oil from the propeller to allow the propeller to go to feather.

master engine The engine which sets the reference RPM for a synchronization system.

oil pump The governor component which generates oil pressure at a value high enough to allow constant-speed propeller operation.

onspeed The constant-speed governor condition where the system is operating at the selected RPM and the force of the speeder spring is balanced by the centrifugal force on the flyweights. The pilot valve in the governor is positioned to trap oil in the propeller.

overspeed The constant-speed governor condition where the system is operating an RPM higher than that set by the speeder spring. The centrifugal force on the flyweights is sufficient to overcome the force of the speeder spring, raise the pilot valve, and direct oil to increase the propeller blade angle.

overspeed governor A governor which is designed to increase the propeller blade angle if the RPM exceeds the system rated RPM by a slight amount.

pilot valve The valve located in the middle of the governor drive shaft which is used to direct oil pressure to and from the propeller during constant-speed operation. The position of the pilot valve is determined by the forces exerted by the flyweights and speeder spring in the governor.

propeller The aerodynamic device which is designed to convert engine power and RPM to useful thrust.

relief valve A type of check valve which is used to regulate governor oil pressure at a constant value.

reset valve A component of an overspeed governor which resets the governor to a lower RPM to allow a ground check of the governor's operation.

RPM The rotational speed of a system (engine) measured in revolutions per minute.

sensing unit The component of the synchronization system which is used to measure the RPM of an engine. The signal generated by each sensing unit is fed to the comparison unit.

slave engine The engine in a synchronization system which has its RPM adjusted to the same RPM as the master engine.

slip ring The ring on the rear of some propellers which is used to return the Beta control valve to the neutral position as the position of the propeller piston changes.

speeder rack The component on the head of the governor which is moved in and out of the governor by the pilot's control lever. This component adjusts the compression of the speeder spring.

speeder spring The spring in the head of the governor between the speeder rack and flyweights which opposes the movements upward of the pilot valve by the flyweights.

sump The engine lubrication system oil reservoir.

synchronization Having all engines operate at the same RPM.

toe The lower portion of a flyweight which is in contact with the pilot valve and applies the lifting force to the pilot valve.

underspeed The constant-speed governor condition where the system is operating at an RPM lower than that set by the speeder spring. The centrifugal force on the flyweights is reduced, allowing the speeder spring to cause the pilot valve to lower, and direct oil to decrease the propeller blade angle.

unfeather To decrease the propeller blade angle to some angle lower than feather to restart the engine.

variable force A force, such as oil pressure, which is increased or decreased to control a change in propeller blade angle. This force is opposed by the fixed force.

windmill The autorotation of the propeller by the action of the air force striking the forward blade surface when the engine is not producing power.

ANSWERS TO STUDY QUESTIONS

Section I

1. Engine RPM
2. Culver Cadet
3. Provisions for governor drive and internal oil tranfer seals.
4. Hartzell

Section II

1. Two to three hundred psi
2. Inlet side of the governor pump
3. Direct oil to, or release oil from, the propeller
4. It is lowered
5. Speeder spring
6. Rearward
7. Allow feathering of the propeller

Section III

1. They are balanced
2. Tilt inward
3. Toe
4. Increase
5. Move throttle forward
6. Move propeller control forward
7. Decrease in blade angle
8. From the propeller

Section IV

1. Governor oil pressure
2. Lift rod
3. Latch mechanism
4. 100 psi
5. Move the cabin propeller control forward
6. Engine oil sump

Section V

1. Grittiness or binding
2. Cold oil or preservative oil
3. A release agent
4. Toward the governor

5. The high RPM stop screw

6. 1/8-inch

8. Aircraft Type Certificate Data Sheet and Supplemental Type Certificates

Section VI

1. Check for evidence of oil leaks

2. To check for components movement when the engine is developing thrust and binding which may be hidden by vibration

3. Replace the governor with a known good governor

4. Air coming out of the engine crankcase breather line or the oil filler

5. A blockage breaking loose and causing damage or injury

6. Congealed oil

7. Unfeather electric switch

Section VII

1. To maintain all engines at the same RPM

2. RPM sensing units, comparison unit, and actuator unit

3. Segments on the flyweight plate or cup moving past a magnetic pickup

4. Master engine

5. To prevent excessive overspeeding by increasing propeller blade angle

6. To allow a ground check of the overspeed governor

7. Throttle

8. Beta control valve

Aircraft Governors Final Examination

Student________________________________

Grade________________________________

Place a circle around the letter for the correct answer to each of the following questions.

1. With a governor controlled propeller system, what happens to the propeller blade angle as the aircraft accelerates during the takeoff?

   A. Blade angle increases.

   B. Blade angle decreases.

   C. The blade angle does not change.

   D. The blade angle may increase or decrease depending on the particular system.

2. The middle section of a governor is normally referred to as the:

   A. Head.

   B. Body.

   C. Base.

   D. Cover.

3. The oil pressure developed by the governor oil pump is:

   A. Regulated by the pilot valve.

   B. Always directed to the propeller oil chamber.

   C. Varied as the RPM varies.

   D. Held constant by the governor relief valve.

4. As RPM increases, the governor flyweights will:

   A. Tilt inward.

   B. Become rigid vertically.

   C. Tilt outward.

   D. Engage from the drive shaft.

5. Concerning a feathering system, when the cabin control for the propeller is moved to the feather position:

   A. The governor drive shaft disengages from the engine.

   B. The pilot valve is lowered.

   C. The lift rod is lowered.

   D. The pilot valve is raised.

6. If the flyweights of the governor are tilting inward, the system is:

   A. Overspeed.

   B. Onspeed.

   C. Underspeed.

   D. Feathering.

7. If the throttle is increased during flight:

   A. A pilot valve will be lowered and blade angle will decrease.

   B. The pilot valve will be raised and the blade angle will increase.

   C. The pilot valvé will be lowered and the blade angle increase.

   D. A pilot valve will be raised and the blade angle will decrease.

8. What mechanism may be used to prevent a feathering propeller from feathering when the engine is stopped on the ground at the end of a flight?

   A. Latch mechanism.

   B. Counterweight mechanism.

   C. Accumulator mechanism.

   D. Spring mechanism.

9. A propeller accumulator is normally located in the:

   A. Engine compartment.

   B. Landing gear wheel well.

   C. In the passenger compartment.

   D. Aft of the passenger compartment.

10. When an engine is running normally, the air pressure in an accumulator is equal to:

    A. Engine oil pressure.

    B. The ground service air charge in the accumulator.

    C. Propeller oil pressure.

    D. Governor oil pressure.

11. What is the source of unfeathering oil pressure in a crossfeed system?

    A. The operating engine's governor oil pump.

    B. The operating engine's oil pump.

    C. The electrically operational oil pump.

    D. The hydraulically operational oil pump.

12. Which of the following governor mounting gaskets incorporates a filter screen?

    A. Shipping gasket.

    B. Run-in gasket.

    C. Governor mounting gasket.

    D. All of the above.

13. One revolution of the governor high RPM adjustment screw is equal to about:

    A. 10 RPM.

    B. 25 RPM.

    C. 50 RPM.

    D. 100 RPM.

14. Engine RPM for a synchronization system is normally sensed by the:

    A. Governor oil pump pressure gage.

    B. Propeller control lever position.

    C. Engine tachometer.

    D. Propeller governor.

15. How many slave engines are used in a 4-engine aircraft?

A. 1

B. 2

C. 3

D. 4

16. Synchronization is not normally used during:

A. Takeoff.

B. Climb.

C. Cruise.

D. Descent.

17. An overspeed governor is:

A. Controllable during normal flight by the cabin control.

B. Required on all aircraft.

C. Prevents the propeller from feathering.

D. Set for one RPM during flight.

18. Beta Mode operation refers to operations during:

A. Ground movement.

B. Takeoff.

C. Cruising flight.

D. Descent.

19. The Beta control valve:

A. Is operated by the cabin propeller control.

B. Is operated by the cabin throttle control.

C. Is mechanically linked to the pilot valve.

D. Is used to feather the propeller.

20. The propeller slip ring:

A. Rotates with the propeller.

B. Is directly controlled by the cabin propeller control.

C. Is directly controlled by the cabin throttle control.

D. Is used as part of the synchronization system.

Aircraft Governers

Answers to Final Examination

1. A
2. B
3. D
4. C
5. D
6. C
7. B
8. A
9. A
10. D
11. A
12. C
13. B
14. D
15. C
16. A
17. D
18. A
19. B
20. A

NOTES

NOTES

NOTES